DC FLUX PARAMETRON

A New Approach to Josephson Junction Logic

World Scientific Series in Computer Science

Volume 1: Computer-aided Specification Techniques
by J Demetrovics, E Knuth & P Radó

Volume 2: Proceedings of the 2nd RIKEN International Symposium on Symbolic and Algebraic Computation by Computers
edited by N Inada & T Soma

Volume 3: Computational Studies of the Most Frequent Chinese Words and Sounds
by Ching Y Suen

Volume 4: Understanding and Learning Statistics by Computer
by M C K Yang & D H Robinson

Volume 5: Information Control Problems in Manufacturing Automation
by L Nemes

Volume 6: D C Flux Parametron
A new approach to Josephson junction logic
by E Goto & K F Loe

Volume 7: Advances in Syntactic and Structural Pattern Recognition
edited by H Bunke & A Sanfeliu

Series in Computer Science — Vol. 6

DC FLUX PARAMETRON

A New Approach to Josephson Junction Logic

K F Loe
National University of Singapore

E Goto
University of Tokyo and RIKEN

Published by

World Scientific Publishing Co Pte Ltd.
P. O. Box 128, Farrer Road, Singapore 9128
242, Cherry Street, Philadelphia PA 19106-1906, USA

Library of Congress Cataloging-in-Publication Data

Goto, Eiichi, 1931-
  DC flux parametron.

  (Series in computer science; vol. 6)
  1. Parametrons. 2. Josephson junctions. 3. Logic circuits.
I. Loe, K. F. (Kia Fock) II. Title. III. Series
TK7895.P3G67   1986   621.39'5   86-15724
ISBN 9971-50-113-9

Copyright © 1986 by World Scientific Publishing Co Pte Ltd.

All rights reserved. This book, or parts thereof, may not be reproduced in any form or by any means, electronic or mechanical, including photocopying, recording or any information storage and retrieval system now known or to be invented, without written permission from the Publisher.

Printed in Singapore by Kyodo-Shing Loong Printing Industries Pte Ltd.

FOREWORD

Parametron was invented about thirty years ago when the transistor was in the infancy stage of development. During that time, it was not known for sure whether the transistor or the parametron would turn out to be the better logic components for building computers. In fact, a parametron consisting of passive electrical components such as resistors, inductors and capacitors, was more reliable and economical to produce at that time. However with the breakthrough of the semiconductor technology, the transistor soon became the more reliable and economical device for making computer components. Since then the transistor continues as the main device in making computers.

Notwithstanding the remarkable progress, semiconductor technology is almost reaching its physical limit in terms of speed and integration. Thus the search for new technology is imperative for our insatiable greed for computer power. Looking around at this moment, it seems that there are not many alternatives in the technology horizon that hold the promise to provide technology for the future needs of computer power. The dc flux parametron using Josephson junction and operating on the principle of parametron seems to be a promising alternative since it combines the very high switching speed and the highly integratable Josephson junction with the established parametron computer technology.

This book was conceived for two important reasons which are mutually relevant. First, recently some important results, both theoretically and experimentally, were obtained in this area of research. Second, the significances of these results can only be better understood if we know the background of the parametron computer technology.

Therefore the book not only provides the basic background knowledge of dc flux parametron but also the recent works being done in this area and the potential applications of this dc flux parametron for the high speed computer design. Also two papers in the Appendix about the historical development of this technology are given. We hope it would help the readers to understand the works which have been done leading to this development.

One of the authors, Loe Kia Fock, who was a foreign research scientist of the University of Tokyo and the RIKEN (Institute of Physical and Chemical Research) when the manuscript of this book was prepared, would like to thank the researchers and the staff there for their assistance in many aspects.

Recently, the dc flux parametron was experimentally confirmed to operate at 1.8 GHz clock. The results will be published in the *IEEE Transaction on Magnetics*.

CONTENTS

FOREWORD v

Chapter 1. INTRODUCTION
1.1 The Needs for Novel Technology 3
1.2 The Josephson Computer Technology 4
1.3 Scope of Study in this Book 5

Chapter 2. REVIEW OF JOSEPHSON COMPUTER TECHNOLOGY
2.1 A Josephson Tunnel Junction 10
2.2 Dynamics in the Josephson Junction 13
2.3 Conventional Josephson Circuit Devices 14
2.4 The Earliest Josephson Circuit Device 17
2.5 Multijunction Josephson Devices 18
2.6 Current Injection Josephson Device 20
2.7 Introduction to Josephson Parametron Devices 22

Chapter 3. PARAMETRON COMPUTER TECHNOLOGY
3.1 Basic Parametron Circuit and Principle 27
3.2 Basic Parametron Logic Circuits 32
3.3 Parametron Computer Logic Circuits 44
3.4 Esaki Diode High-Speed Logic Circuit 47
3.5 Principles of Josephson Parametron Devices 52

Chapter 4. CIRCUIT THEORY BASED ON MECHANICAL APPROACHES
4.1 Formulations of Lagrangian and Hamiltonian 59
4.2 An Example of Lagrangian and Hamiltonian 61
4.3 Mechanical Formulations of Circuitry 63
4.4 Examples of Lagrangian Approach to Josephson Circuitry . 67
4.5 Examples of Hamiltonian Approach to Josephson Circuitry 71

Chapter 5. ANALYSIS OF DIRECT CURRENT FLUX PARAMETRON
5.1 Basic Formulation of DCFP 76
5.2 Operational Principle of DCFP 79
5.3 Static Characteristics of DCFP 81

5.4	Input and Output Correlations	83
5.5	Logic Loading and Higher Mode Hazards	85
5.6	Leakage Inductances in Transformer Coupling	87
5.7	DCFP with Leakage Inductances	89
5.8	The Supercurrent Noise	90
5.9	Reduction of Supercurrent Noise by Split Excitations	92
5.10	Thermal Noise at Transition State	95
5.11	Quantum Tunnelling and Thermal Noise	96
5.12	Simulations and Numerical Analysis	99
5.13	Experiments on DCFP Circuits	104
5.14	Measurements of High Speed DCFP Operations	104

Chapter 6. CIRCUIT SIMULATION CODE GENERATOR

6.1	Design Motivations and Characteristics of CSCG	114
6.2	Principles and Criterions for CSCG Design	116
6.3	Algebra Algorithms for CSCG	118
6.4	Implementation of Modular Design and Specifications	123
6.5	Implementation of the Complete Specifications	128
6.6	Output and Documentation	130
6.7	Examples of CSCG Applications	132

Chapter 7. ANALYSIS OF INDUCTIVE JOSEPHSON LOGIC (I JL)

7.1	Potential Energy and Current of I JL	144
7.2	Formulations of General Form of I JL Circuit	148
7.3	Examples of I JL Circuits	151
7.4	Incorporating I JL Modules into CSCG for Simulations	154

Chapter 8. COMPUTER TECHNOLOGY BASED ON DCFPs

8.1	Constraints and Wiring Rules	161
8.2	Impossibility of Directional Coupling	163
8.3	Relay Noise	165
8.4	Applications of DCFP to Memory and CPU Designs	173
8.5	Conclusions	175

REFERENCES 177

APPENDIX
1. The Parametron, a Digital Computing Element which Utilizes Parametric Oscillation
   E Goto 183
2. Esaki Diode High-Speed Logical Circuits
   E Goto *et al.* 196

Chapter 1
INTRODUCTION

From its initial discovery in 1911 it took almost 50 years for the origin and properties of superconductivity to be satisfactorily explained by Bardeen, Cooper and Schrieffer<Bardeen57> in their now famous BCS theory of superconductivity. This theory assumes that certain electrons coupled into paired state(Cooper pair) which is responsible for all the characteristics of superconductivity. Based on the theory of BCS, Josephson predicted<Josephson62> in 1962 that interactions of electrical potential and magnetic fields with the Cooper pair should be observable in the so called weakly coupled superconducting regions, and the relationship of dc current over the junction to be dependent on the magnetic flux were experimentally verified at the Bell Telephone Laboratories in 1963<Anderson63>. Subsequently the superconductive tunnel junction, combined with a means of controlling the magnitude of the zero-voltage current to form a fast, low-power logic or a memory device was recognized in 1967 by Matisoo<Matisoo67>. His idea was eventually developed into a large scale research in IBM Research in Yorktown, Zurich, and East Fishkill as well as other research institutes.

However in the last few years the idea of zero-voltage current controlled Josephson device was abandoned by most of the major research institutes owing partly to the emerging of new ideas of flux controlled Josephson logic devices <Likharev76, Likharev77>, which dissipate even less power with the better integration and performance. In this book a new Josephson logic device using entirely fluxes for input, output and clocking control, and based on the parametron computer technology<Goto59> to be called direct current flux parametron or DCFP, will be studied. Some of the study and idea given here are also applicable to other flux controlled devices.

1.1 The Needs for Novel Technology

Over the past 20 years, the technology of computer systems has advanced drastically in terms of performance, cost and reliability. There is every reason to expect this advance to continue, at a rate almost as fast as we have experienced to date. However, the advance already achieved has pushed the mechanisms of switching, storage and communication close enough to the fundamental physical limits to bring into awareness for the first time the limitation in the engineering of high-performance systems.

Fundamental limits such as the permissible size of the transistors on a silicon chip to be reduced by another factor of 10 (but probably not a factor of 100), although in principle, would lead to an order of magnitude improvement in computer performance, mundane problems, such as how to provide wiring to interconnect these smaller devices, could prevent us from realizing much of this potential improvement. The three basic functions required in computing systems are: (1) switching that is non-linearity and amplification in addition to logical operation such as provided by a transistor, (2) storage of information in electronic form that is a variation in stored energy, such as the quantity of charge stored on a capacitor that must be large enough to assure reliability, and (3) communication of information that is bringing operands into proximity, normally accomplished by some form of wire that unavoidably interposes several kinds of degradation of signals, including delay noises.

The needs for computer power have pushed the conventional technology to its physical limit in coping with the requirements of the above three basic functions in computing systems, and the search for new alternative, which may provide the better ultimate performance, is becoming imperative. From the view point of performance, Josephson technology is perhaps the ultimate. The advantages, which are the characteristics of the technology, will be given in the following.

1.2 The Josephson Computer Technology

Josephson devices are attractive for ultrahigh-performance computers because of the combination of three characteristics: (1) extremely fast switching speed, (2) extremely low power dissipation, and (3) operation at very low temperature (around $4°K$).

The first characteristic of fast switching speed is clearly a prerequisite for an ultrahigh-performance computer, the switching speed of a Josephson junction is in the order of less than $10psec$ using bridge type Josephson junction, however in this book clocking signal of $100psec$ is used for circuit simulation, based on the tunnelling type of Josephson junction which operates in a speed about one order slower than the bridge type junction but it is easily available in large quantity now.

The second characteristic of low power dissipation offers advantages since it allows the transfer of heat directly from chips into the liquid coolant bath. This is in contrast with the need for voluminous heat sinks necessary in current high-performance semiconductor technology to prevent heat runaway. On the other hand the zero-voltage current controlled Josephson circuit device has the power dissipation in the order of $500nW$ per circuit, and the flux controlled Josephson junction device to be discussed in this paper has a lower power dissipation of only $1nW$ for a clocking signal of $100psec$ with critical junction current of $100\mu A$. Thus heat transfer problem in our approach will even be less significant and higher integration is possible.

The third characteristic of low operating temperature provides very desirable features for ultrahigh-performance computers, since the low temperature allows strip lines interconnecting the chips to be made to become superconducting(both the line and ground plane), thus eliminating resistance losses that distort and attenuate high speed pulses travelling down nonsuperconductivity lines. Because of the zero resistance, the superconducting line can be made narrower, and this allows the lines to be placed closer together, permitting the circuit packing density to be increased. Low temperature can provide even more benefits; the thermal energy (kT) is about 100 times smaller at $4°K$ than at 360 to $380°K$ where semiconductor circuits are operated. This leads to smaller thermally induced electrical noise, and helps to maintain good signal-to-noise ratio, in addition the thermally activated processes that can lead to reliability problems in the package and chips can also greatly be reduced at the temperature of $4°K$.

1.3 Scope of Study in this Book

In this book a new type of Josephson computer technology named dc flux parametron(DCFP) computer technology will be studied. Besides the difference in logic principle, there is a fundamental difference between our approach and the conventional approach since in our approach a logic state is represented by a flux polarity whereas the conventional approach uses a voltage state to represent a logic state. Chapter 2 will review the conventional Josephson Junction logic and some of the comtemporary Josephson devices. An introduction to the DCFP will also be given in this chapter. Chapter 3 will discuss the parametron principle and the logics of parametron. It is interesting to point out that recently IBM has proposed a Josephson device<Zappe75> which works in the principle similar to the original parametron whereas the DCFP to be studied here is similar to the Esaki diode high speed logic circuit which is an improved device from parametron. The merits of DCFP as compared with the IBM Josephson parametron will be discussed after review of the conventional parametron and Esaki diode high speed logic circuit.

In chapter 4 mechanical approach to analyse the Josephson junction circuit will be introduced. The dynamic of this mechanical approach is developed using the Lagrangian and the Hamiltonian formulations. Based on the mechanical approach, in chapter 5, the DCFP operational principle will be analysed. The works include finding the proper ranges of the circuit parameters and studying the characteristics of DCFP by taking into consideration the logic loading problem, higher mode operational problem, noise problem and tunnelling problem. Finally simulations were done with the proper parameter values to confirm the analytical predictions.

When using DCFP to design basic computer logic circuits, a simulation system was conceived to assist the design works as to be discussed in chapter 6. The differences of our simulation system as compared with the current conventional circuit simulation systems are that the

Hamiltonian is used to formulate the circuit dynamics and computer algebra is used to manipulate the algorithms and to generate the simulation FORTRAN codes. Thus this system is called Circuit Simulation Code Generator(CSCG). The system is currently in use for our design works. In chapter 7, study about Inductive Josephson Logic(IJL) circuits will be introduced. IJL circuits seem to be a set of useful logic circuits. They can couple with DCFP to perform some useful logic functions. Only the basic idea of IJL will be studied in this chapter; experiments and ways to use the IJL circuits shall be pursued further. Based on the simulation results we shall discuss in chapter 8 the design rules and constraints when coupling DCFPs together to make logic circuits. The constraints in the fan-in and fan-out will be discussed. The relay noise in DCFP circuit couplings will be studied. Methods to alleviate the relay noise will be proposed. Finally the DCFP computer architecture currently under study<Shimizu84> will be brought out for discussion to substantiate the viability and potential usage of the DCFP computer technology.

Chapter 2
REVIEW OF JOSEPHSON COMPUTER TECHNOLOGY

In this chapter we shall review the conventional Josephson computer technology. What we mean by conventional Josephson computer technology is the so called zero-voltage current control type of logic device pioneered by Matisoo<Matisoo67>. His idea was eventually developed into an active research throughout the world in many research laboratories. Computer technology such as logic devices, memory units, packaging techniques, fabrication and functional units were developed surrounding this idea. Amongst all the research groups IBM Research in Yorktown, Zurich, and East Fishkill probably had done the most works on these aspects and came out with various kinds of logic devices.

Recently new approaches to make Josephson devices, which hold promises for the better performance in terms of speed, power consumption and circuit integration, are emerging. The main theme of study in this book is an approach proposed by E.Goto<Goto84> using completely flux for consideration in circuit operations and logic state representations. Before going into the detail study of this new approach it is worthwhile to review the conventional approach in order to understand the significant differences of this new approach. There are many excellent review papers on the conventional Josephson computer technology<IBM80, Zappe83> we shall only capture the essential aspects which reflect the main differences from our study.

2.1 A Josephson Tunnel Junction

The nonlinear characteristic of the Josephson junction is used as the active element on which all present superconducting digital circuits are based. In this section the fundamental electrical properties will be reviewed.

The boson-like Cooper pairs that carry the supercurrent behave cooperatively and minimize the total energy by locking their phases. As a result, superconductivity can be described as a single entity of pairs that is identified by a Schrodinger-like wave function. If the superconductor carries current, the pairs have a net momentum and, although the wave functions remain locked, there is a gradient of phase<VanDuzer81>. The phase difference φ between two points along a superconducting path is an important parameter, related to the current carried by the superconductor.

Assuming now that two superconductors, S_1 and S_2, as shown in Fig.2.1(a), are weakly coupled through a thin tunneling barrier of few Angstrons with λ_1 and λ_2 representing the superconducting penetration depths. Such a structure is a Josephson junction. A Josephson point-like junction carries a current determined by the phase difference φ as,

$$I = I_m \sin \varphi \qquad (2.1)$$

and a voltage across the junction is given by,

$$V = K\frac{d\varphi}{dt} \qquad (2.2)$$

Finally,

$$\text{grad}\varphi = D(\mu_0 H \text{x} n)/K \qquad (2.3)$$

where I_m is the zero field Josephson threshold current, $K=\hbar/2e$, $D=(\lambda_1+\lambda_2+\delta)$, H the external magnetic field, μ_0 the permeability of free space and n the unit vector perpendicular to the plane of the junction. The above three are the Josephson equations<Josephson62>.

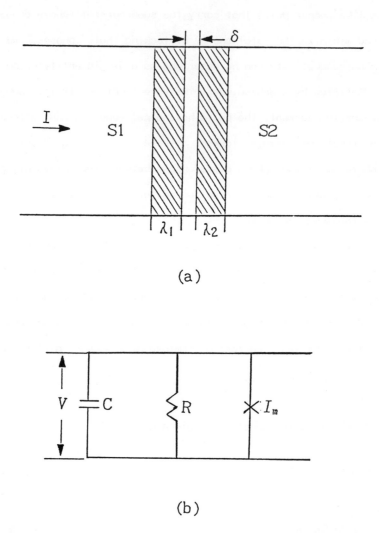

Fig.2.1 (a) Section through a Josephson tunnel junction formed between two superconductors S_1 and S_2. If its dimensions are small compared to the Josephson penetration depth, it is a point junction that can be described by a simple equivalent circuit as shown in (b).

2.2 Dynamics in the Josephson Junction

The current density of a junction $j_1 = I_m/a$, where a is the junction area, is a strong exponential function of the barrier thickness. This indicates that care must be taken in growing the tunnel barrier for practical devices in which current levels should not change by more than a few tens of percent. In contrast, within any given design range, the junction capacitance C is only a weak function of j_1 and can to first order be related to an assumed constant specific capacitance C_s as $C \sim C_s a$. The equivalent circuit of a junction is also shown in Fig.2.1(b). It has a capacitance C, a non-linear quasiparticle tunneling resistance $R(V)$ and a Josephson current element as specified by equation(2.1). By adding all the currents one obtains,

$$C\frac{dV}{dt} + \frac{V}{R} + I_m \sin\varphi = I \qquad (2.4)$$

and by substituting V as defined in equation(2.2) one gets,

$$CK\frac{d^2\varphi}{dt^2} + \frac{Kd\varphi}{Rdt} + I_m \sin\varphi = I \qquad (2.5)$$

Equation(2.5) defines the dynamics of a Josephson point-like junction, a junction smaller than the Josephson penetration depth λ_J, in which magnetic self fields can be neglected. Also, magnetic stray fields which penetrate the junction are assumed much smaller than the magnetic flux quantum $2\pi K = h/2e = \Phi_0 = 2.07 \times 10^{-15} Vs$, so that for any practical purpose I_0 can be assumed constant. The Josephson penetration depth is given by,

$$\lambda_J = \sqrt{\Phi_0 / 2\pi\mu_0 D j_1} \qquad (2.6)$$

Equation (2.5) was simultaneously proposed by Stewart<Stewart68>, and by McCumber<McCumber68>, who showed its equivalence to the equation of motion of a pendulum. This was extremely fruitful to the field since mental interpretations were possible in terms of a mechanical analog. Another mechanical model is that of the washboard in which a particle moves on a tilted sinusoidal potential<VanDuzer81>. Both models were widely used to illustrate the dynamic phenomena in the conventional Josephson computer circuit device.

2.3 Conventional Josephson Circuit Devices

There were a number of ways in which conventional switching devices were made from the tunnel junction. In each case, however, the operating principles are the same. All devices exhibit an I-V characteristic similar to that of Fig.2.2. The two states of the device are the zero-voltage state and the resistive state. In operation, the device is current biased in the zero-voltage state with $I_g < I_m(0)$ as shown in Fig.2.3, and is caused to switch under the influence of an input (or control) current I_c, which either adds to I_g or reduces $I_m(0)$ so that the threshold is exceeded, and the device switches according to the external load to the resistive state. Switching from the resistive state to the zero-voltage state can occur in one of the two ways. Either the device switches to the zero-voltage state upon the removal of the input (operation referred to as nonlatching), or by reducing the bias current I_g such that the voltage across the device becomes less than a characteristic voltage V_{min}. Either device operation can be obtained by appropriate choice of device and circuit parameters. A thorough discussion of the relevant considerations for this latching

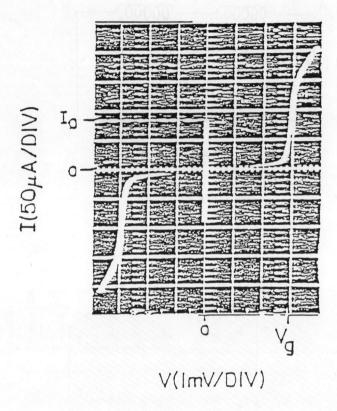

Fig.2.2 *I-V* characteristic of a Josephson tunnel junction. It has two states, the zero-voltage state up to I_0 and the voltage state as indicated by the non-linear single particle tunnelling curve.

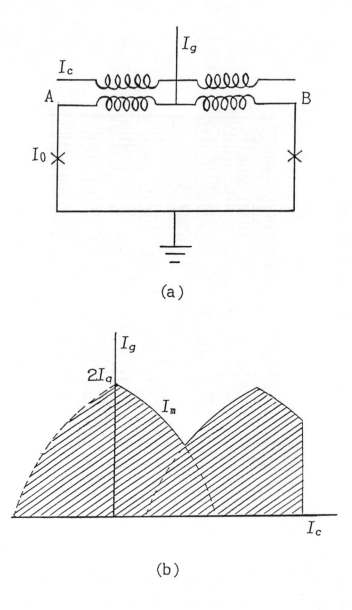

Fig.2.3 (a) is a typical zero-voltage controlled device, where I_g is the biasing current and I_c is the input control current which either adds to I_g or reduces I_{m0} so that two states zero-voltage which is shown in the shaded area of (b), and voltage state which is outside the shaded area of (b) can be obtained.

and nonlatching operation was given by Zappe<Zappe75a>. Nonlatching operation can be obtained only for very small load impedance, consequently most of the logic circuit are latching.

A very significant design feature of the device is the threshold characteristic, the locus of points in the I_g and magnetic field plane which forms the boundary between the $V=0$ and $V\neq0$ states of the device. It is essentially this feature and the operating current levels which form the distinction between various classes of device. The conventional Josephson junction devices can be classified into three types as discussed in the following sections.

2.4 The Earliest Josephson Circuit Device

The first and oldest type of these devices is the in-line gate with single or multiple controls<Matisoo67>. This device was in use as the basic logic and the memory device from its inception in 1965 until superseded by multijunction devices in 1974. The device structure consists of a tunnel junction with an overlaid in-line control, or set of controls, which is coupled by magnetic field to the junction, thus providing control of the zero-voltage current. To obtain current gain(loosely defined as the current supplied to the load, divided by the input current), the device current levels have to be sufficiently large that this requirement restricts the devices for optimally driving relatively low-output impedances.

As the technology has evolved from $25\text{-}\mu m$ minimum linewidths to the current $2.5\text{-}\mu m$ linewidth, the output impedance have increased, leading to incompatibility with the in-line device properties and multijunction devices were introduced. In-line devices still find use, however, in applications in which large current levels are necessary, for example, as drive and logic devices in the DRO main memory chip <Gueret80>.

2.5 Multijunction Josephson Devices

The second class of device consists of the multijunction devices, which came into being in respond to the need for device structures whose threshold currents were sensitive to smaller levels. It was known through the work of Jaklevic et al <Jalklervic65> that this could be achieved by interconnecting two point junctions with a junction devices carrying the acronym "dc SQUID" (dc superconducting quantum interference device) and has been used extensively as sensing elements of uniquely high-sensitivity instruments, magnetometers, graiometers, gradiometers and voltmeters. To make practical switching devices the magnetic field is coupled via an overlaying control line as in the in-line gate.

The two device structures of primary interest are the center-feed two-junction device and the split-feed three-junction structure, in which the point junctions have a zero-voltage current ratio of 1:2:1. The threshold characteristics of these devices are shown in Figs.2.4(a) & (b), respectively.

The threshold characteristic of the two-junction SQUID of Fig.2.4(a) shows overlapping modes. Under the central lobe, no circulating current is present in the loop formed by the two point junctions and the interconnecting inductances, whereas in the two adjacent lobes a circulating clockwise or counterclockwise current, corresponding to a flux quantum, is present in addition to the externally applied

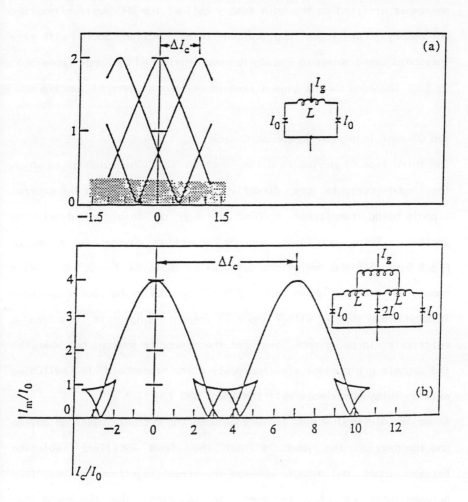

Fig.2.4 Threshold characteristics for multijunction devices: (a) two-junction device, $\Phi_0/LI_0=0.98$; (b) three-junction device, $\Phi_0=LI_0=7$.

currents. The fact that the modes overlap about the origin means that $+\Phi_0$ and $-\Phi_0$ can be stored in the two-junction SQUID with no bias. This device is utilized as the main memory cell of the DRO(destructive read out)memory. For logic application three-junction SQUID with the threshold curve shown in Fig.2.4(b) was introduced by Zappe<Zappe75b>. In fact the idea can be generalized to devices of several junctions.

2.6 Current Injection Josephson Device

The third type of device is a two-junction SQUID<Gheewala80a> in which the input currents are directly injected, rather than the control signals being transformer coupled through an insulated overlaying control. This nonlinear current injection device is shown in Fig.2.5(a) with the device characteristics shown in Fig.2.5(b). With inputs A and B as shown in Fig.2.5(a), the device basically operates in such a way that if either input is present singly it requires a relatively large signal level for the device to switch, but when the two signals are present simultaneously a low threshold to switching exists. This approximate well the ideal AND function.

A major distinguishing feature between the current injection device and the previous two types is that they have excellent isolation between input and output, whereas in current injection the isolation between input and output is poor. In addition, the threshold for determining the sensitivity is solely determined by the switching of the non-zero voltage state of a single Josephson junction so that only a current gain of at most 1 may be obtained. Therefore, although it is advantageous to use it as a switch, it is difficult to apply it to various kinds of logic circuits. Consequently, the injection devices

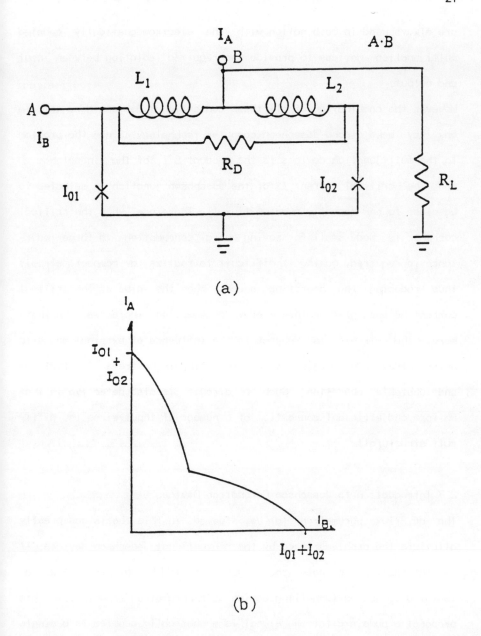

Fig.2.5 (a) An equivalent circuit of a current injection gate. (b) The threshold curve of the gate.

are always used in combination with the electromagnetically coupled multijunction devices to provide the required isolation between input and output.

However the combination of all these conventional devices does not in any way make a good Josephson computer technology. Since the problem in the multijunction devices is the product $L.I_j$ of the inductance L and the critical current I_j of the Josephson junction is selected to be close to one magnetic quantum flux Φ_0. Therefore, when the critical current is made small for saving energy consumption, a large inductance is required, making it difficult to realize a compact circuit thus reducing the operating speed. When the value of the critical current becomes greater then energy consumption increases. Furthermore, the circuit is subject to the influence of external magnetic noise, stray inductance and so on, resulting in extreme fluctuations and unstable operation. Such a circuit is also defective in that uniform and efficient connection of a number of input wires is difficult structurally.

2.7 Introduction to Josephson Parametron Devices

The dc flux parametron to be studied in this thesis can greatly alleviate the problems faced by the conventional Josephson devices. It is interesting to note that Zappe of IBM has proposed to use an isolated Josephson tunnelling device as a parametron<Zappe76>. In his proposal a pump(excitation) signal is magnetically coupled to a single junction Josephson device which functions as a parametron. The pump signal amplifies the signal in the parametron in several periods of oscillation. The two phases of subharmonic oscillations in the parametron are being used to designate the two boolean logic states. This idea mimics the original parametron device invented by E.Goto<Goto59>. Recently E.Goto proposed a dc flux parametron(DCFP) using a pair of Josephson junction. His idea is similar to the Esaki diode high speed logical circuit <Goto60>. Esaki diode high speed logical circuit is the improved idea from the original parametron though the basic logic principle is the same for both. We shall discuss more details of Zappe's Josephson parametron device and the DCFP in the next chapter after we have reviewed the principles of parametron and the basic idea of Esaki diode high speed logical circuit.

Chapter 3
PARAMETRON COMPUTER TECHNOLOGY

Parametron was invented by E.Goto in 1954<Goto59> using capacitors, ferrite-core coils and resistors as the basic elements to build a device which generates subharmonic oscillation by an oscillating pump signal. Subharmonic oscillator was known prior to the invention of parametron, however the subtle of the parametron idea is to take advantage of the two possible phases of the subharmonic oscillator signal as the logic states for computers. When operating with a two-to-one division ratio, the subharmonic oscillator can have two distinct signal phases, 0 and π. These two phases can represent "0" and "1" in a binary digital computer.

The parametron was utilized as a majority logic element, so that when any odd number of inputs was used, one phase would be predominate and the parametron would lock and amplify that phase. Together with the NOT function which was simply achieved by an inverting transformer, the majority logic element suffices for all logical operation such as flip-flop, binary counter, binary full adder and parity circuit. It can easily be shown that AND and OR are special cases of majority logic where one of the three inputs are biased by signals representing "0" or "1" as to be illustrated later.

To control the phase of the parametron, a quench mechanism is needed to destroy the existing information. The parametron is then re-energized and the growing subharmonic locks in phase with the input signal. If no input signal is present, the phase is undeterminate and is initiated by noise in the circuit. Computer logic circuits generally require a clock and parametron computers are no exception, however it requires clocking for a reason even more basic than timing.

Once the parametron is oscillating, energy travels bidirectionally from it, feeding back into the input as well as into the output. Two phases clock was not suffice to achieve a unidirectional information flow, thus three phase clock system was used for this purpose, and was utilized in all the parametron computers. Each parametron is quenched for approximately half the clock period and is active for the other half. Of the three clock phases, the two successive phases overlap so as to couple signal in the proper direction through the circuit.

For reliability reason, three inputs are often used in the majority logic in the parametron circuit though five is occasionally used. Owing to the high gain obtainable from the parametron, fanout of 10 to 20 are possible.

3.1 Basic Parametron Circuit and Principle

Parametron is essentially a resonant circuit in which either the inductance or the capacitance is made to vary periodically. Fig.3.1 shows circuit diagram for parametron elements which consists of coils wound around two magnetic ferrite toroidal cores F_1 and F_2, a capacitor C, and a damping resistor R, and a small toroidal transformer T. Each of the cores F_1 and F_2 has two windings and these are connected together in a balanced configuration, one winding $L=L'+L''$ forming a resonant circuit with the capacitor C and being tuned to frequency f. An exciting current, which is a superposition of dc bias and a radio frequency current of frequency $2f$, is applied to the other winding, $l'+l''$, causing periodic variation in the inductance $L=L'+L''$ of the resonant circuit at frequency $2f$.

The operation of the parametron is based on a spontaneous generation

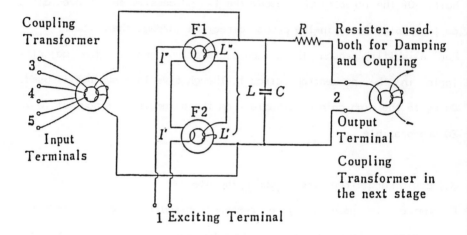

Fig.3.1 A parametron consists of coils wound around two magnetic ferrite toroidal cores F1 and F2, a capacitor C, and a damping resistro R and a small toroidal transformer T.

of a second-subharmonic parametric oscillation, that is a self-starting oscillation of frequency f, in the resonant circuit. Parametric oscillation is usually treated and explained in terms of Matheiu's equation. A more intuitive explanation may be obtained by the following consideration.

Let the inductance L of the resonant circuit be varied as,

$$L = L_0(1 + 2\Gamma \sin 2\omega t) \qquad (3.1)$$

where $\omega=2\pi f$, and let us assume the presence of a sinusoidal ac current I_f in the resonant circuit at frequency f, which can be broken down into two components as follows:

$$I_f = I_s\sin(\omega t) + I_c\cos(\omega t). \qquad (3.2)$$

Then, assuming that the rate of the variation of amplitudes of the sine and cosine components, I_s and I_c are small compared with ω, the induced voltage V will be given by,

$$V = \frac{d}{dt}LI_f = \omega L_0(I_s\cos \omega t - I_c\sin \omega t) + 3\Gamma\omega L_0(I_s\sin 3\omega t + I_c\cos 3\omega t) + \Gamma\omega L_0(-I_s\sin \omega t + I_c\cos \omega t) \qquad (3.3)$$

The first term shows the voltage due to a constant inductance L_0, and the second term or the third harmonic term may be neglected since it is off resonance. The third term, which is essential for the generation of the second subharmonic, shows that the variable part of the inductance behaves like a negative resistance $-r=-\Gamma\omega L_0$ for the sine component I_s, but behaves like a positive resistance $+r=\Gamma\omega L_0$ for the cosine component I_c

Therefore, when the circuit is nearly tuned to ω the sine component I_s (A in Fig.3.2) of any small oscillation in it will build up exponentially (B in Fig.3.2) while the cosine component damps out

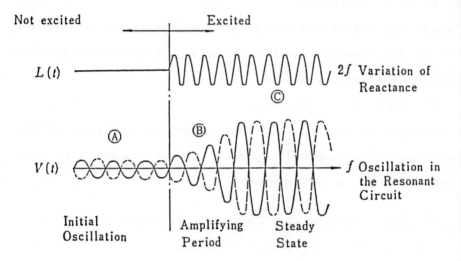

Fig.3.2 Oscillation of Parametron, where small oscillation A is exponentially building up in B and reaching a stationary state of C.

rapidly. If the circuit were exactly linear, the oscillation would increase indefinitely. Actually, the non-linear characteristic of the core causes deturning and increase of hysteresis loss, so that a stationary state (C in Fig.3.2) is rapidly reached. However, in contrast to the ordinary oscillators, the phase of stationary oscillation of a parametron has a definite relation to the exciting current $I_{2\omega}$, and there are two possible modes differing by π radian to each other in phase. These two modes of oscillation are respectively shown by the solid line and the dotted line in Fig.3.2. Which mode of stationary oscillation (C) is realized is determined by the sign of the sine component of small initial oscillation (A). An initial oscillation of quite small amplitude is sufficient to control the mode or the phase of stationary oscillation which is to be used as the output signal. Hence, the parametron has an amplifying action.

The existence of dual modes of stationary oscillation can be made use of to represent a binary digit "0" and "1" in a digital system, and thus a parametron can store one bit of information. However, oscillation of parametrons in this stationary state is extremely stable, and if one should try to change the state of an oscillating parametron from one mode to another just by directly applying a control voltage to the resonant circuit, a signal source as powerful as the parametron itself would be necessary. This difficulty can be surmounted by providing a means for quenching the oscillation, and making the choice between the two modes, *i.e.*, the rewriting of information, by a weak control voltage applied at the beginning of each building up period. Actually, this is done by modulating the exciting wave by a periodic square wave which also serves as the clock pulse of the computer.

Hence, for each parametron there is an alternation of active and passive periods, corresponding to the switching on and switching off of the exciting current. Usually, the parametron device uses three clock waves, labeled I, II and III, all having the same pulse recurrence frequency, but switched on and off one after another in a cyclic manner as shown in Fig.3.3. This method of exciting each of the parametrons in a digital system with either one of the three exciting waves I, II and III is usually called the "three beat" or the "three subclock" excitation.

3.2 Basic Parametron Logic Circuits

Since a digital system of any complexity can be synthesized by combining the four basic circuit elements, namely DELAY, AND, OR and NOT, it will be seen that a complete digital computer system can, in principle, be constructed using only one kind of active circuit element--the parametron, which has two phases of oscillation difference by π radian. In the following we shall look at these basic logic circuits.

(a) *Majority Logic Circuit*

Suppose that parametron has three terminals 1, 2 and 3, each of which receives input current i_1, i_2 and i_3 equally of frequency ω, as shown in Fig.3.4. In the event that i_1 and i_2 of these inputs currents are of 0 phase, and the remaining i_3 is of π phase, these currents may be expressed by the following equations:

$$i_1 = I_1 \qquad i_2 = I_2 \qquad i_3 = -I_3 \tag{3.3}$$

where I_1, I_2 and I_3 denote absolute values of i_1, i_2 and i_3 respectively. Since these currents are all combined to flow into tuning circuit of parametron, resulting current i_0 flowing into the tuning

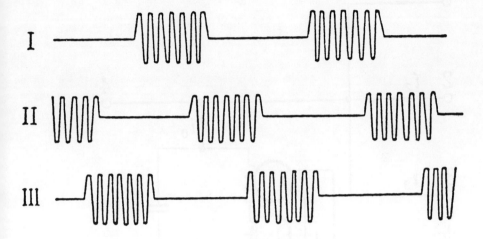

Fig.3.3 The exciting currents of three subclocks namely phases I, II and III excitations.

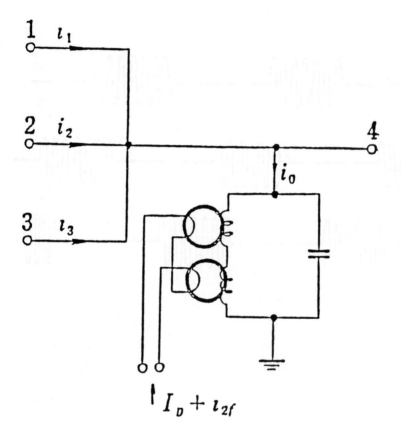

Fig.3.4 Illustration on majority decision selection (majority logic) with three inputs i_1, i_2, i_3 and one output i_0.

circuit becomes,

$$i_0 = i_1 + i_2 + i_3 = I_1 + I_2 - I_3 \qquad (3.4)$$

Now if the three currents have the same absolute input values denoted by,

$$I_1 = I_2 = I_3 = I, \qquad (3.5)$$

then equation (3.4) becomes,

$$i_0 = I(1+1-1) = I.$$

which is equal to a 0 phase current. Thus, when 0 phase or π phase is expressed as +1 or -1, respectively, the combined current may be obtained as the current corresponding to a phase which is found from the sum of these numbers. Thus the oscillation phase is determined by the majority decision in all the input currents.

It is obvious that this majority decision principle may be applicable to any odd number of inputs, in most of the application three inputs with occasional five inputs are used. So far, we have assumes that all the input currents have the equal absolute values. In actual case, however, the currents flowing into the respective input terminals are supplied from parametron located at the previous stage, and consequently they can not be considered to have the same absolute values, nevertheless their differences are not so large because they are limited to definite values by the amplitude shaping function of the coils in the parametron as described earlier.

(b) *AND Logic Circuit*

Consider a parametron with three input terminals x, y and c, and an output terminal z as shown in Fig.3.5(a), and assume that the terminal c to be always fed with a current of phase (-1) and of absolute value equal to the input currents at terminals x and y. Then, for the various

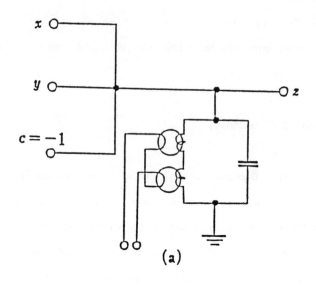

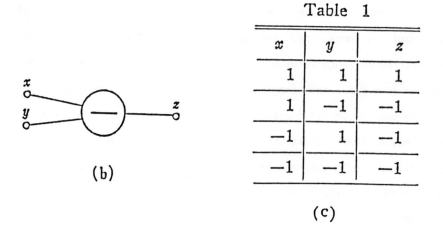

Fig.3.5 (a) Majority logic circuit used as an AND circuit, (b) symbolized AND circuit, (c) tabulating of the inputs to the circuit.

possible combinations of inputs from terminals x and y, in terms of 0 phase (+1) or π phase (-1), the phase of output terminal z expressed in terms of the aforementioned +1 or -1 may be derived from the majority decision principle with $c=-1$, in the following manner:

(i) for $x=1$ and $y=1$ then $z=x+y+z=1+1-1=1$,

(ii) for $x=1$ and $y=-1$ then $z=-1$,

(iii) for $x=-1$ and $y=1$ then $z=-1$,

(iv) for $x=-1$ and $y=-1$ then $z=-3$.

The -3 in (iv) indicates that the input current i_0 becomes $-3I$, that is a current three times the I at π phase. Though the absolute value of the resultant input i_0 is larger as compared with the earlier two cases, z corresponding to oscillation output may still be denoted as $z=1$, because the output is made constant after limitation of the oscillation amplitude to a constant value irrespective of absolute value of input current. This may be tabulated as shown in Table 1.

The table shows that output z becomes 1 (0 phase) if and only if both x and y inputs are 1 (0 phase). If x and/or y become -1 (π phase), then the output z becomes -1 (π phase), then the output z becomes -1 (π phase). Such an AND circuit is symbolized as shown in Fig.3.5(b) indicating that terminal c is always applied with -1 (π phase).

(c) *OR Logic Circuit*

Consider the parametron with three input terminals x, y and c, and an output terminal z, as shown in Fig.3.6(a), and assume that the terminal c to be fed with a current of 0 phase (+) in contrast with the aforementioned AND circuit, then following the similar reasoning we obtained Table 2 as shown in Fig.3.6(c)

The table shows that output z becomes -1 (π phase) only if both inputs

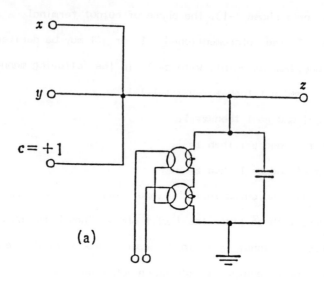

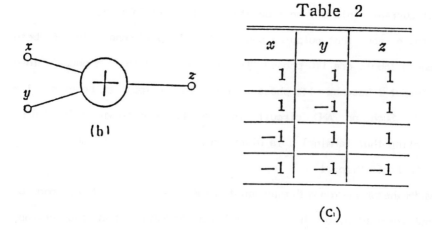

Fig.3.6 (a) Majority logic circuit used as an OR circuit, (b) symbolized OR circuit, (c) tabulating of the inputs to the circuit.

x and y are $-1(\pi$ phase). If x and/or y are $+1(0$ phase), then the output z becomes $+1(0$ phase). Such an OR circuit is symbolized as shown in Fig.3.6(b), indicating that terminal c is always applied with $+1(0$ phase).

(d) *DELAY Circuit*

To make it possible to use parametron as a binary circuit, it must be able to transmit the information of binary numbers. The transmission route in this case is also usable as a delay circuit. Such a delay circuit or transmission circuit for information of binary numbers used is shown in Fig.3.7(a) or(b). In the former figure parametrons are connected by an impedance Z, while in the latter figure parametrons are connected by an impedance Z and a transformer. A parametron delay line is symbolized as in Fig.3.7(c)

Now we shall describe the operation of this circuit taking the case of parametron II in Fig.3.7.

Let oscillation voltages of I,II and III be denoted by E_I, E_{II} and E_{III}, and the respective excitation currents by I_e, II_e and III_e. Now we assume that parametron I oscillates with 0 phase, producing oscillation voltage of +1. Through a coupling resistance, this oscillation voltage gives parametron II a small voltage of 0 phase, though the parametron II does not oscillate so long as it is not applied with excitation. But if the parametron II is applied with excitation II_e during application of the excitation on I, II is oscillated with 0 phase due to coupling of 0 phase from the parametron I. After complete growth of the oscillation, excitation current I_e of parametron I is interrupted as shown in Fig.3.8. As the oscillation of parametron I stops and it looks as if 0 phase oscillation of parametron I were

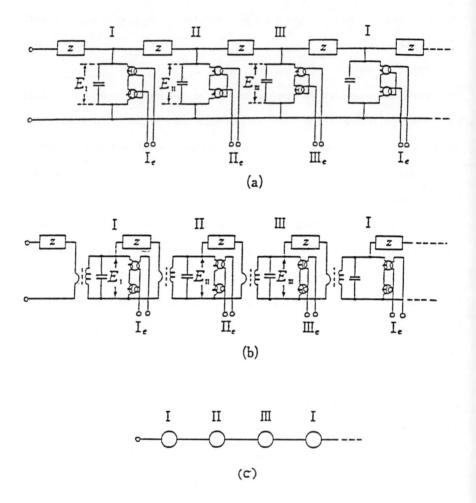

Fig.3.7 (a) parametron is connected by an impedance Z, (b) parametron is connected by a transformer and an impedance Z, (c) Symbolized parametron delay circuit.

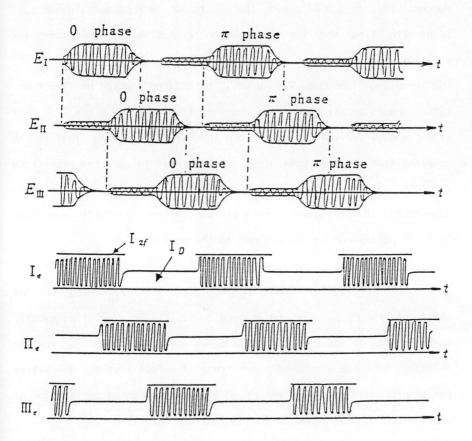

Fig.3.8 Quenching pattern of excitation and oscillation currents in a parametron.

transferred to parametron II. The successive similar excitation in I, II and III sequence which overlap with each other respectively in parts of their excited times, allow the informations to be sent successively. Fig.3.8 shown the excitation current-oscillation patterns with time and the information transmission mechanisms for oscillation phase.

The coupling impedance Z is usually a resistance which has two functions: one is to strengthen the coupling from previous stage so that it becomes sufficiently greater than the skipped coupling, that is the coupling from the next post stage: the other is to make the rise time constant of oscillation equal to the delay time constant by properly adjusting Q of the tuning circuit of oscillation so that quenching speed of parametron may be brought to the maximum.

The transformer used is not directly required for a delay circuit, but is provided for the later mentioned NOT circuit for reversing current phase of the input. In the delay circuit, the delay of time exactly corresponding to one quenching period (the keying period of 2ω) may be obtained by three stages of parametrons. The fact that the delay time can be prolonged to any extent by varying the quenching period is of great important not found in the conventional circuits.

(e) *NOT Circuit*

In the pulse circuit, a "NOT circuit" is defined as a circuit which makes conversion between mark and space. In the parametron circuit, however, the "NOT circuit" is defined as a circuit which converses +1 (0 phase) to -1 (π phase) or -1 (π phase) to +1 (0 phase), and therefore, an inverted transformer which is shown in Fig.3.9(a) may be said a NOT circuit. This circuit symbolized as shown in Fig.3.9(b).

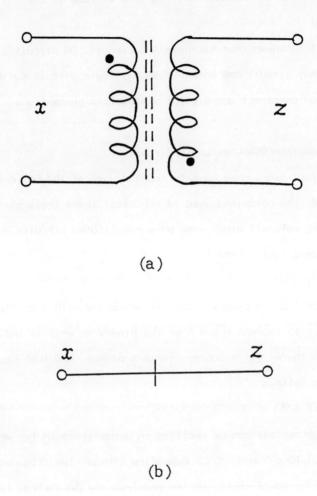

Fig.3.9 (a) an inversed transformer NOT circuit, (b) Symbolized NOT circuit.

Relation between input x and output z in this circuit, corresponding to a negation in mathematical logic as express in the following form:

$$z = \bar{x} \qquad (3.6)$$

Since it is known that if these AND circuit, OR circuit, NOT circuit and delay circuit are known, then any logic circuit can be made, that is we can construct any digital system from parametrons.

3.3 Parametron Computer Logic Circuits

In the last section we have introduced some of the basic logic circuits and the notations used to represent these logic circuits, in the following we shall study some more complicated circuits synthesis from these basic logic circuits.

(a) *Half Adder*

Since addition in binary system is processed with a carrier bit, it is necessary to express a sum S of the binary as well as indicating that whether there is a carry forward or not, and such expressions are given as follows:

$$S = X\bar{Y} + \bar{X}Y = (X + Y)\overline{XY} \quad C = XY \qquad (3.6)$$

These expressions can be realized in parametrons in two ways as shown in Fig.3.10(a) and (b). Comparing these two figures shows that figure(b) circuit configuration required one parametron less than (a).

(b) *Flip-flop Circuit*

Fig.3.11 shows a parametron flip-flop or a 1-bit memory circuit. Three parametrons, coupled in ring form, are required to store 1 bit of information. In Fig.3.11 (a) and (b) it is assumed that the signals in the set and reset inputs are both normally "0". The flip-flop will be set to "1" when a "1" signal is applied to the set input, and the

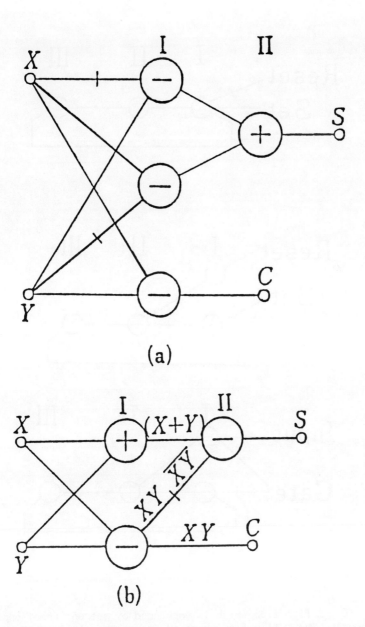

Fig.3.10 (a) a half adder with four parametrons, (b) a half adder with three parametrons.

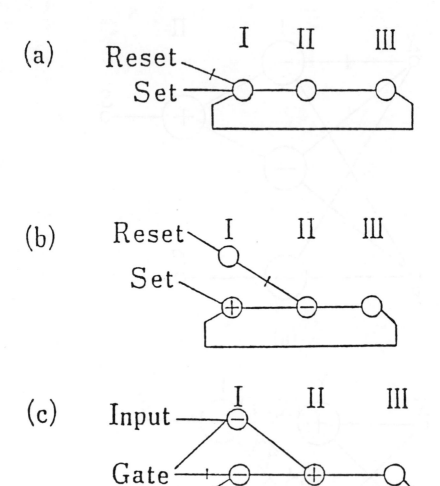

Fig.3.11 (a) A flip-flop which is unchanged by set and reset applied simultaneously, (b) a flip-flop which is reset by applied simultaneously set and reset, (c) a flip-flop with an input gate.

flip-flop will be reset to "0" when a "1" signal is applied to the reset input. The functional difference between Fig3.10(a) and (b) consists in that, when both the set and the reset signal are applied simultaneously, the stored information will not change in the circuit of Fig.3.11(a), but it will be reset to "0" in the circuit of Fig.3.11(b). Fig.3.11(c) shows a flip-flop with a gate. As long as "0" is applied to the gate, the stored information does not change, but when "1" is applied to the gate, the signal from the input is transferred to the flip-flop.

(c) *Binary Counter*

Fig.3.12 shows three stages of binary counting circuits connected in cascade, thus forming a scale-of-8 counter. Three flip-flops are included in this circuit to store a 3-bit count. In the quiescent state, in which "0" is applied to the input, the bits stored in each flip-flop do not change, but each time a "1" is applied to the input for a single clock period, the registered binary number is increased by 1 (mod 8).

(d) *Binary Full Adder*

Binary full adder circuit may be obtained by combining two half adders so as to add the carry to the next higher order, however a simpler circuit can be constructed for this purpose as shown in Fig.3.13.

Various circuits actually used in the parametron computer construction can be found in references<Takahasi60>.

3.4 Esaki Diode High-Speed Logic Circuit.

An Esaki diode is a two-terminal negative resistance element which is essentially bilateral. Therefore, unlike ordinary transistor switching

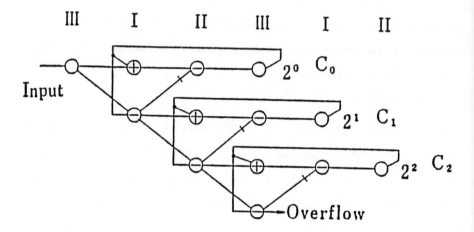

Fig.3.12 Three stages of Binary counters forming a scale of 8 circuits.

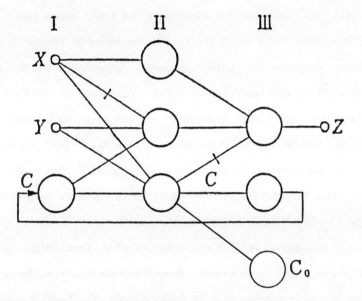

Fig.3.13 A full adder circuit with carry saving loop.

circuits, Esaki diode circuits require that some special method to be incorporated for obtaining a unilateral characteristic for the transmission and amplification of digital signals. This situation is completely analogous to the parametron which we have studied earlier. However there is a basic difference in these circuits due to the different ways to control signals. In parametron the exciting clock and the input/output signals are ac in nature, whereas the exciting clock and the input/output signals of the Esaki diode logic circuit are dc in nature, thus there are some differences in realizing some of the logic circuits in both approaches, especially the NOT circuit which can not be realized by a inverted transformer in the Esaki diode logic circuit but a more elaborated way was invented<Goto60>. Nevertheless all the methods used to construct the flip-flop, counter and etc., for parametron circuits are applicable to Esaki diode logic circuit, thus in the following we shall only briefly explained the basic principle of Esaki diode logic circuit.

A typical voltage-current characteristics of an Esaki diode is illustrated in Fig.3.14(a). It clearly shows the negative resistance A and B which is the characteristic of Esaki diodes. Two Esaki diodes which have almost the same charactieristic can be connected in series to form the basic circuit as shown in Fig.3.14(b), which was called "twin circuit"(or Goto pair). When a input seed signal of binary values is fed to the middle point of this twin circuit, then a dc clock signal shown in Fig.3.14(c) is used to excite and amplify the input circuit as shown in Fig.3.14(c). Two polarities of amplified voltage signals, which can be used to designate two logic states, are thus obtained.

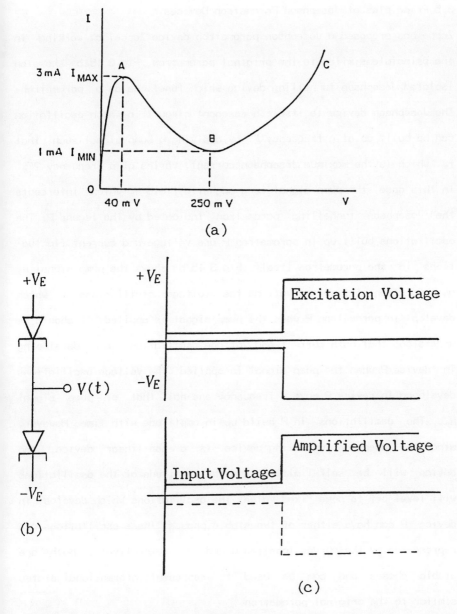

Fig.3.14 (a) A typical characteristic of a silicon Esaki diode, (b) Esaki diode pair used as logic circuit, (c) the voltage excitation characteristics of the voltage polarities.

3.5 Principles of Josephson Parametron Devices

Zappe has proposed a Josephson parametron device<Zappe75c> working in the principle similar to the original parametron. Fig.3.15(a) shows an isolated Josephson tunnelling device which functions as a parametron. The Josephson device is itself a resonant circuit in which oscillation can be built up at a frequency F_r by applying a pump signal such that I_m, which is the maximum Josephson current, varies at a frequency $2F_r$. In this case, the pump signal is a magnetic field H_p which intercepts the Josephson tunnelling parametron, indicated by the legend P. The oscillations built up in parametron P are voltage and current fluctuations in the parametron itself. Fig.3.15(b) shows the pump signal H_p as a function of time, as well as the voltage oscillations V which develop in parametron P when the pump signal is applied. It should be understood that both current and voltage oscillations are developing in device P when the pump signal is applied. The voltage oscillations developed in device P have a frequency one-half that of pump signal H_p. The oscillations in P build up in amplitude with time. However, since the Josephson tunnelling device is a non-linear device, the device will be self-limiting and the amplitude of the oscillations will level off to some fixed value. The oscillations which develop in device P can have either of two stable phases. These oscillations are represented by the curves labelled 0 and π, respectively. Both are stable phases and can be used to represent informational states similar to the original parametron.

In this book the main theme of study is another Josephson parametron device known as a DCFP which operates in a way more similar to the

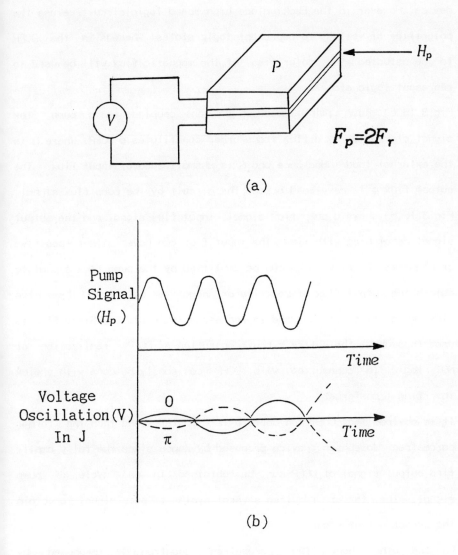

Fig.3.15 (a) A Josephson junction parametron circuit, (b) the voltage excitation characteristic of phase polarities.

Esaki diode high speed logic circuit than the original parametron device. However in the Esaki diode high speed logic circuit we use the polarities of voltage to represent logic states, whereas in the DCFP to be studied the polarities of the magnetic flux will be used to represent logic states.

Fig.3.16(a) shows a pair of Josephson junctions coupled to a pump flux signal (clock) Φ_e by a flux transformer constitutes a DCFP. Where L is the external load inductance and S is a small seed of input flux. The output flux Φ is developed across the circuit by the pump flux signal. Fig.3.16(b) shows a pump flux signal, input flux signal and the output signal developing with time. The input flux can have either positive or negative polarity which can be amplified by the pump flux signal to obtain the output flux of positive or negative polarity. These two flux polarities can be used to represent logic states. Since flux is used throughout the entire circuit operation of DCFP, realization of NOT logic in connection with DCFP can easily be done with a flux inverting transformer.

It is obvious that DCFP can operate in a faster speed than the former parametron Josephson device proposed by Zappe since the fully amplified output signal of DCFP can be obtained in one cycle of pump signal. The former required several cycles of pump signal to obtain the proper output signal.

On the other hand the parametron quantron (PQ) proposed by Likharev<Likharev77> using single Josephson junction shares more similarities to the DCFP, therefore some of the results obtained here should also be valid in the PQ.

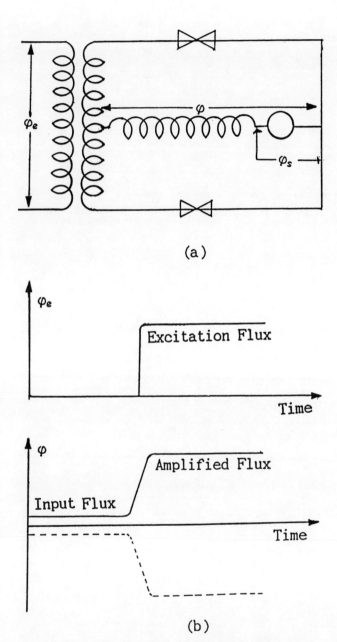

Fig.3.16 (a) A dc flux parametron made from a pair of Josephson junctions, (b) the flux excitation characteristic of the flux polarities.

Chapter 4
CIRCUIT THEORY BASED ON MECHANICAL APPROACHES

In fact, the foundation for the circuit concept was laid in the early development of the theory of dynamics. When Newton linked force to acceleration in the relation

$$\frac{d}{dt}(mv) = F \qquad (4.1)$$

known as the Newton's second law, and d'Alembert expounded his principle<Karman-40>:

"that every state of motion may be considered at any instant as a state of equilibrium of appropriate inertia forces are introduced", we had the basic elements of the circuit concept available. We could set at once, for a single particle at any instant, the externally applied force equal to the sum of the "inertia force", of the resistive force depending on velocity, and of the restoring force depending on position, and evolve a "circuit representation".

However, the development of physical thought and of mathematical expression are not always concurrent. As far as mechanics was concerned, many practical problems, particularly those involving rotation, led to the search for more general formulations of the equations of motion. The most powerful formulations of the equations of motion are the Lagrangian formulation and the Hamiltonian formulation. On the other hand, the circuit theory being commonly used in the present day is mainly based on the considerations of conservative of charge and current-voltage relation of the circuit devices. In this book we shall return to adopt the mechanical approaches for circuit dynamic. The reason that we adopt these approaches is because a mental interpretation of the flux behaviour in the Josephson device is possible with a mechanical view, and quantum nature of flux can be studied using mechanical approach. Also a circuit simulation system was developed based on Hamiltonian approach as to be given later.

4.1 Formulations of Lagrangian and Hamiltonian

We consider a classical system with f degrees of freedom. For N point particles, f will be equal to $3N$. We suppose that we have a set of generalized coordinates for the system:

$q_1, q_2, \ldots, q_f$.

These may be Cartesian, polar, or some other convenient set of coordinates. The generalized velocities associated with these coordinates are

$\dot{q}_1, \dot{q}_2, \ldots, \dot{q}_f$

The expression of Newton's second law by the Lagrangian equation of motion is

$$\frac{d}{dt}(\frac{\partial \mathcal{L}}{\partial \dot{q}_i}) - \frac{\partial \mathcal{L}}{\partial q_i} + \frac{\partial \mathcal{D}}{\partial \dot{q}_i} = F_{ext} \qquad (i=1,2,\ldots,f), \qquad (4.2)$$

where for a simple non-relativistic system the Lagrangian $\mathcal{L}$ is given by

$$\mathcal{L}(q_i, \dot{q}_i) = K - U. \qquad (4.3)$$

Here K is the kinetic energy and U the potential energy of the system. Equation (4.2) can easily be verified if q_i are Cartesian coordinates, for then we have

$$\mathcal{L} = \frac{1}{2}\sum_j M_j \dot{q}_j^2 - U \qquad (4.4)$$

and, letting $q_i = x$,

$$M\ddot{x} = -\frac{\partial V}{\partial x};\qquad(4.5)$$

but $-\frac{\partial V}{\partial x}$ is just the x component of the force F, and we have simply

$$F_x = M\ddot{x}.\qquad(4.6)$$

The Hamiltonian form of the equations of motion replaces the f second-order differential equations (4.2) by $2f$ first-order differential equations. We define the canonical momenta by

$$p_i = \frac{\partial \mathcal{L}}{\partial \dot{q}_i}.\qquad(4.7)$$

The Hamiltonian $\mathcal{H}$ then is defined as

$$\mathcal{H}(p_i, q_i) = K_p + U,\qquad(4.8)$$

where K_p is the kinetic energy in terms of canonical momenta derivable from the kinetic energy K in (4.3) using (4.7). The Hamiltonian form of equations of motion is given by,

$$\frac{dq_i}{dt} = \frac{\partial \mathcal{H}}{\partial p_i}\qquad(4.9)$$

and

$$\frac{dp_i}{dt} = -\frac{\partial \mathcal{H}}{\partial q_i} - \left(\frac{\partial D}{\partial \dot{q}_i}\right)_p,\qquad(4.10)$$

where $\left(\frac{\partial D}{\partial \dot{q}_i}\right)_p$ is a function of canonical momenta.

From the mechanics point of view, a Lagrangian gives a space-time evolution of a mechanics system, and a Hamiltonian is the total energy of a system. The two ways to describe a mechanics system is equivalent, however in some situation we may find that one way will provide a more convenient way to express or to handle a problem than the other. For instance, when we analyse the resonant characteristics of the DCFP using mechanical approach, the equations of motion derived from the Lagrangian approach would provide a convenient way to interpret the DCFP resonant circuit and to define some of useful parameters for it. On the other hand, when we develop a circuit simulator system for DCFP study, we found that Hamiltonian approach would be much convenience, as it will provide a easier way to design the interface and program algorithms. As we go along in this book the application of both approaches will be explained further.

In comparison with the Newton second law, Lagrangian and Hamiltonian approaches are less familiar to many people, therefore as the first example to illustrate both the approaches, we shall look at a harmonic oscillation system.

4.2 An Example of Lagrangian and Hamiltonian

Fig.4.1 shows a harmonic oscillator with dissipation. The mass of the particle is m, the restitutional force is $-kx$ and the friction is $-\beta\dot{x}$, where we used the Cartesian coordinate x to denote the displacement of the particle, and the canonical momentum is denoted by p. The potential energy U, the kinetic energy K and the dissipating function D are respectively given as follows,

$$U = \frac{k}{2}x^2, \qquad (4.11)$$

$$K = \frac{m}{2}\dot{x}^2 \qquad (4.12)$$

$$D = \frac{\beta}{2}\dot{x}^2 \qquad (4.13)$$

The Lagrangian $\mathcal{L}$ is,

$$\mathcal{L} = K - U$$

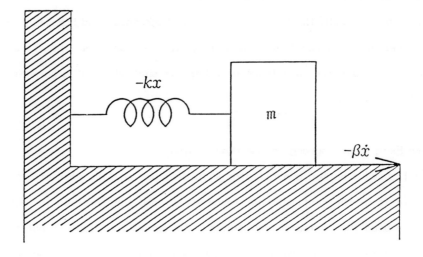

Fig.4.1 A mechanical damped oscillating system consists of a particle with mass m.

$$= \frac{m}{2}\dot{x}^2 - \frac{k}{2}x^2 \qquad (4.14)$$

The Lagrangian equation of motion is by (4.2) as

$$m\ddot{x} + \beta\dot{x} + kx = 0 \qquad (4.15)$$

To obtain the Hamiltonian we shall find the canonical momentum first. Canonical momentum is given by

$$p = \frac{\partial \mathcal{L}}{\partial \dot{x}} = m\dot{x} \qquad (4.15)$$

then K_p and $(\frac{\partial D}{\partial \dot{x}})_p$ are obtainable as follows:

$$K_p = \frac{1}{2m}p^2 \qquad (4.16)$$

and

$$(\frac{\partial D}{\partial \dot{x}})_p = \frac{\beta}{m}p \qquad (4.17)$$

Therefore,

$$\mathcal{H} = K_p + U$$

$$= \frac{1}{2m}p^2 + \frac{k}{2}x^2. \qquad (4.18)$$

According to (4.9) and (4.10) the Hamiltonian equations of motion are,

$$\frac{dx}{dt} = \frac{1}{m}p \qquad (4.19)$$

and

$$\frac{dp}{dt} = -kx - \frac{\beta}{m}p. \qquad (4.20)$$

4.3 Mechanical Formulations of Circuitry

As we have mentioned in our introduction to this chapter that circuit concept was originally laid on the foundation of dynamics. Description of linear electric circuits was in terms of the dynamics of particles because of the greater familiarity with mechanical systems. The most important and actually decisive difference was the fact that in electrical systems we separate the three basic energy namely kinetic energy, potential energy and dissipating energy localizations in individual physical elements. It is therefore natural to introduce individual symbols for these elements and indeed, any graphical diagram of connections is also a pictorial schematic of the physical arrangement, quite in contrast to the mechanical system.

In most of the circuit analysis, voltage and current are used as variables to describe the system operations, however the device DCFP in this book will be studied based on flux and current. Therefore if we are to make use of the Lagrangian or Hamiltonian approach to deal with the DCFP, we shall convert voltages to fluxes for all the circuit components. Define 'coordinate' φ which is, in fact, flux variables as follows,

$$\int v dt = \varphi \tag{4.21}$$

Fig.4.2 shows a damped RCL oscillator. In term of the flux variable the magnetic energy stored in the coil or inductance of this circuit is given by

$$U = \frac{1}{2} L i_L^2 = \frac{1}{2L}(\int v dt)^2 = \frac{1}{2L}\varphi^2 \tag{4.22}$$

The electric field energy stored in the condenser or capacitor is given by

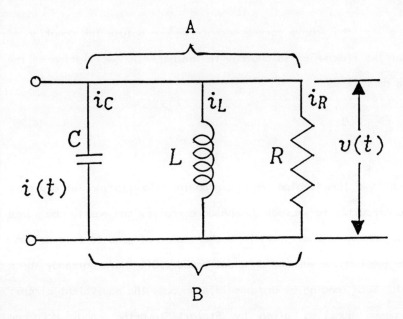

Fig.4.2 A electrical damped oscillating system consists of RCL components.

$$K = \frac{C}{2}v^2 = \frac{C}{2}\dot{\varphi}^2 \qquad (4.23)$$

The dissipating electric energy through the resistor is given by

$$D = \frac{1}{2R}v^2 = \frac{1}{2R}\dot{\varphi}^2 \qquad (4.24)$$

According to the above correspondence we can obtain the equation of motion of the flux which is analogy to the harmonic oscillation of the particle as follows,

$$C\ddot{\varphi} + \frac{\dot{\varphi}}{R} + \frac{\varphi}{L} = 0 \qquad (4.25)$$

Extending the Hamiltonian and Lagrangian description of ordinary electric circuitry to include Josephson circuitry can easily be done since Josephson circuits with mechanical analogies to to a driven-pendulum model and a washboard model<VanDuzer81> have already been proposed. Such analogies are possible because the equivalent circuit of Josephson junction given by Stewart<Stewart68> and McCumber <McCumber68> is a parallel connection of a resistor, a capacitor and a supercurrent as discussed earlier. The characteristic of the supercurrent can be regarded in some way as a nonlinear inductance which is a function of flux. Fig.2.1(b) shows the equivalent Josephson junction. If Φ_i denotes the flux, Φ_0 denotes the unit quantum flux and I_m denotes the maximum supercurrent in a Josephson junction, then the potential energy, kinetic energy and dissipating function of the junction can be written respectively as follows,

$$U_J = -\frac{\Phi_0 I_m}{2\pi}\cos\left(\frac{2\pi\Phi}{\Phi_0}\right) \qquad (4.26)$$

$$K_J = \frac{C}{2}\dot{\phi}^2 \qquad (4.27)$$

$$D_J = \frac{1}{2R}\dot{\phi}^2 \qquad (4.28)$$

The above expressions are very useful in the subsequent applications to our study of Josephson circuits.

4.4 Examples of Lagrangian Approach to Josephson Circuitry

In the following we shall give two examples to illustrate the applications of Lagrangian approach to describe the dynamic behaviour of the flux in Josephson circuits. There are some differences between the two examples, in the first example all the dynamical variables are independent variables, whereas in the second example one of dynamical variables is a dependent variable. For the latter we can get rid of the dependent variable by either Lagrange multiplier methods or direct substitution. Since all the problems to be encounter in our subsequent study only consist of independent variables, we shall not pursue dependent variable issue further thereafter.

Here and in the subsequent illustration, X_i and x_i will be used to denote flux and phase at some points i of the circuit respectively. Define

$$x_i = 2\pi\frac{X_i}{\Phi_0} \quad I_m = \frac{\Phi_0}{2\pi L_J} \quad E_J = \frac{\Phi_0^2}{4\pi^2 L_J} \quad L_i = A_i L_J \quad t_c = \sqrt{CL_J} \qquad (4.29)$$

where L_i is the loading inductance which is related to the inductance of the Josephson junction L_J by a defined parameter A_i.

To illustrate the applications of the Lagrangian formulation to

Josephson circuits, we shall look at a Josephson circuit shown in Fig.4.3.

For reason of simplicity in our illustration here and in the next example we assume $D = 0$. Thus the potential energy and kinetic energy of the system are,

$$U = \frac{(X_1-X_2)^2}{2L} - \frac{\Phi_0 I_m}{2\pi}(\cos(\frac{2\pi(X_2-X_3)}{\Phi_0}) + \cos(\frac{2\pi X_3}{\Phi_0})) -$$

$$= E_J(\frac{(x_1-x_2)^2}{A} - \cos(x_2-x_3) - \cos x_3) \tag{4.30}$$

$$K = \frac{C}{2}(\dot{X}_2-\dot{X}_3)^2 + \frac{C}{2}\dot{X}_3^2$$

$$= \frac{E_J t_c^2}{2}((\dot{x}_2-\dot{x}_3)^2 + \dot{x}_3^2) \tag{4.31}$$

The Lagrangian equation of motion of this system can be worked out using (4.2) to be,

$$t_c^2(\ddot{x}_2-\ddot{x}_3) - \frac{x_2-x_1}{A} + \sin(x_2-x_3) = 0 \tag{4.31}$$

$$t_c^2(2\ddot{x}_3-\ddot{x}_2) - \sin(x_3-x_2) + \sin x_3 = 0 \tag{4.32}$$

These are differentiate equations of second order, and in general, to find computer algorithms to solve these equations numerically we need to reduce them into a set of first order differentiate equations.

In the next example we shall consider a circuit which illustrates the possible of dependent variables being introduced into a system.

The kinetic energy and the potential energy of the system which is shown in Fig.4.4 can be written as follows,

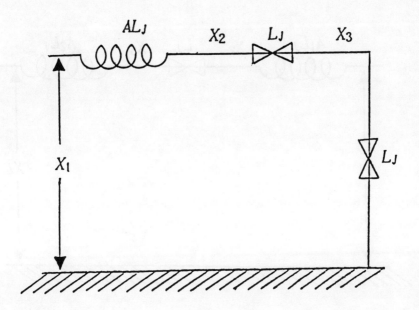

Fig.4.3 A circuit consists of two Josephson junction.

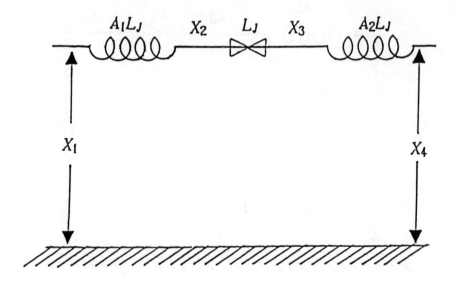

Fig.4.4 A circuit consists of one Josephson junction.

$$U = E_J\left(\frac{(x_1-x_2)^2}{2A_1} + \frac{(x_3-x_4)^2}{2A_2} - \cos(x_2-x_3)\right) \qquad (4.33)$$

$$K = \frac{E_J t_c^2}{2}(\dot{x}_2-\dot{x}_3)^2 \qquad (4.34)$$

The Lagrangian equations of motion are derived by (4.2) as,

$$t_c^2(\ddot{x}_2-\ddot{x}_3) + \frac{x_2-x_1}{A_1} + \sin(x_2-x_3) = 0 \qquad (4.35)$$

$$t_c^2(\ddot{x}_2-\ddot{x}_3) - \frac{x_3-x_4}{A_2} - \sin(x_2-x_3) = 0 \qquad (4.36)$$

From the above expressions we find that the variables of second order differentiation in (4.35) and (4.36) are the same. If we eliminated the second order differentiate variables, we find that one of the variables (e.g. x_2) is a dependent variable of the other (e.g. x_3). It shall be noted that computer algorithms necessary to solve this kind of problems depends on the problem itself. In other words, it is difficult to find a standard algorithm to solve the Josephson circuit problem if the Lagrangian approach be adopted. However in the next section we shall show that if Hamiltonian approach is adopted then both the examples retained the same form of equations of motion after we have eliminated the dependent variable.

4.5 Examples of Hamiltonian Approach to Josephson Circuitry

In the following we shall use the same above two examples to illustrate the applications of Hamiltonian approach to describe the dynamic behaviour of the flux in Josephson circuits. The same conditions and notations will be assumed as in the last section.

In the Hamiltonian approach, unless the kinetic energy is explicitly given in terms of canonical momenta otherwise we need to solve a set of linear equations derived from (4.7) in order to get the canonical form. For the first example of Fig.4.3, we get, from (4.7)

$$\frac{\partial L}{\partial \dot{x}_2} = t_c^2 E_J (\dot{x}_2 - \dot{x}_3) = p_2 \tag{4.37}$$

$$\frac{\partial L}{\partial \dot{x}_3} = t_c^2 E_J (2\dot{x}_3 - \dot{x}_2) = p_3 \tag{4.38}$$

Solving the above simultaneous equations for $\dot{x}_2$ and $\dot{x}_3$ to be in terms of p_2 and p_3, and substitute them into (8) we obtain,

$$K_p = \frac{p_2^2 + (p_2 + p_3)^2}{2 E_J t_c^2} \tag{4.39}$$

According to (4.9) and (4.10) the equations of motion are obtainable as follows,

$$\frac{dx_2}{dt} = -\frac{2p_2 + p_3}{2 E t_c^2} \tag{4.40}$$

$$\frac{dx_3}{dt} = -\frac{p_3}{2 E_J t_c^2} \tag{4.41}$$

$$\frac{dp_2}{dt} = E_J \left(\frac{x_2 - x_1}{A} - \sin (x_2 - x_3) \right) \tag{4.42}$$

$$\frac{dp_3}{dt} = E_J \left(\sin (x_3 - x_2) + \sin x_3 \right) \tag{4.43}$$

This example shows that the equations of motion in the Hamiltonian approach is a set of simultaneous first order differential equations. In the following we shall consider the second example (Fig.4.4) using the Hamiltonian approach.

In the Hamiltonian approach, the canonical variables are obtainable by (4.7) as follows,

$$p_2 = \frac{\partial L}{\partial \dot{x}_2} = E_J t_c^2 (\dot{x}_2 - \dot{x}_3) \tag{4.44}$$

$$p_3 = \frac{\partial L}{\partial \dot{x}_3} = E_J t^2 (\dot{x}_3 - \dot{x}_2) \tag{4.45}$$

eliminating either one of the variables of canonical momenta (e.g. p_3), we obtain the kinetic energy in term of only a canonical momentum as follows,

$$K_p = \frac{p_2^2}{2E_J t_c^2} \tag{4.46}$$

and the Hamiltonian equations of motion are derivable using (4.9) and (4.10) as follows,

$$\frac{dp_2}{dt} = E_J \left(\frac{x_2 - x_1}{A_1} + \sin(x_2 - x_3) \right) \tag{4.47}$$

$$\frac{dp_3}{dt} = E_J \left(\frac{x_3 - x_4}{A_2} + \sin(x_3 - x_2) \right) \tag{4.48}$$

$$\frac{dx_2}{dt} = \frac{dx_3}{dt} = \frac{p_2}{E_J t_c^2} \tag{4.49}$$

The equations of motion still retain the standard form as the first example. Thus it is possible to find a unify algorithms to solve both problems.

Lagrangian and Hamiltonian will be used to study the DCFP in the subsequent chapters. In Chapter 5, Lagrangian will be used to derive the equations of motion for analysing the resonant characteristics of a single DCFP device. When connecting DCFPs together to make various kinds of more complicated logic circuits, the study of these circuit by simulations required a good software simulation tool. In Chapter 6 we shall develop a circuit simulation generator using Hamiltonian approach and computer algebra. The reason that in chapter 5 that Lagrangian method is used is because Lagrangian method provide a familiar form of equations to describe the resonant circuits. In Chapter 6 we find that Hamiltonian approach provide a convenient way to develop a circuit simulation code generator for DCFP circuitry study. In chapter 7 when we study another new kind of logic devices to be called Inductive Josephson Logic(IJL) circuits we shall demonstrate how the circuit simulation code generator system formulated by Hamiltonian approach can easily be extended to incorporate new logic devices into it.

Chapter 5
ANALYSIS OF DIRECT CURRENT FLUX PARAMETRON

In this chapter<Loe84,Loe85>, a new kind of Josephson circuit device known as dc flux parametron or DCFP <Goto84> will be analysed. DCFP will be used as a basic logic device for building computer based on the previous mentioned parametron computer technology. As discussed in chapter 3 that phase or polarity of ac voltage is used in parametron to represent logical signals, whereas polarities of dc flux are used in the DCFP instead. That is how the name DCFP(dc flux parametron) was derived. Three phases clocking and majority logic are used in DCFP in the same way as conventional parametron, and they will not be elaborated here. Also dc voltage is never applied to the Josephson junctions of the DCFP, thus it is distinctly different from the voltage control Josephson device in the conventional Josephson devices. From our study it was found that the tolerance of the super-current deviation in the Josephson junction of DCFP can further be improved by slightly modifying the device. We called the modified DCFP as split excitation DCFP, since the modified device uses double clock one after the other in splitting manner to excite the device.

5.1 Basic Formulation of DCFP

Fig.5.1(a) shows the basic circuit element, where φ_1 and φ_2 are the phase differences over the two identical Josephson junctions, Φ_S is the input flux, Φ_E is the excitation clocking flux and Φ is the output flux.

The potential energy of the system is,

$$U_o = \frac{(\Phi-\Phi_S)^2}{2L} - \frac{I_m \Phi_o}{2\pi}(\cos \varphi_1 + \cos \varphi_2) \tag{5.1}$$

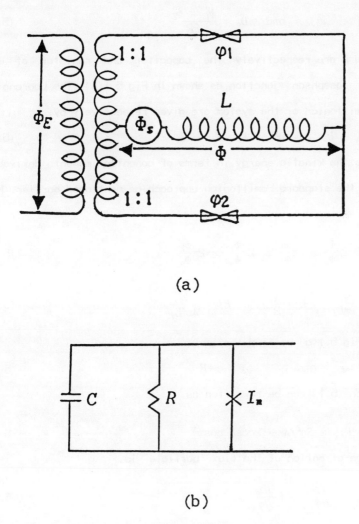

Fig.5.1 (a) A schematic drawing of DCFP, (b) the equivalent circuit of Josephson point like tunnelling junction.

where I_m is the maximum supercurrent and Φ_0 is the unit quantized flux. The kinetic energy and power dissipation function are,

$$K_0 = \frac{C}{2}(\dot{\Phi}_1^2 + \dot{\Phi}_2^2) \quad \text{and} \quad D_0 = \frac{\dot{\Phi}_1^2 + \dot{\Phi}_2^2}{2R} \tag{5.2}$$

where C and R are respectively the capacitor and resistor of the equivalent Josephson junction as shown in Fig.5.1(b). The Lagrangian and the Hamiltonian of the system are given respectively by,

$$\mathcal{L} = K_0 - U_0 \quad \text{and} \quad \mathcal{H} = K_p + U_0 \tag{5.3}$$

where K_p is the kinetic energy in terms of canonical momenta derivable from $\mathcal{L}$ by the standard Hamiltonian approach as mentioned earlier. Now let

$$E = 2\pi\frac{\Phi_E}{\Phi_0} \quad S = 2\pi\frac{\Phi_S}{\Phi_0} \quad \varphi = 2\pi\frac{\Phi}{\Phi_0} \quad \varphi_1 = 2\pi\frac{\Phi_1}{\Phi_0} \quad \varphi_2 = 2\pi\frac{\Phi_2}{\Phi_0} \tag{5.4}$$

and $\quad I_m = \dfrac{\Phi_0}{2\pi L_J} \quad L = \dfrac{A}{2}L_J \quad E_J = \dfrac{\Phi_0^2}{4\pi^2 L_J} \tag{5.5}$

then from Fig.5.1(a) we find that,

$$\varphi_1 = \varphi + E \quad \text{and} \quad \varphi_2 = \varphi - E \tag{5.6}$$

and equation (5.1) can be rewritten as,

$$U_0 = E_J\left((\varphi-S)^2/A - 2\cos E \cos\varphi\right) \tag{5.7}$$

The equation of motion of the flux 'particle' is,

$$\frac{d}{dt}\left(\frac{\partial \mathcal{L}}{\partial \dot{\Phi}}\right) - \frac{\partial \mathcal{L}}{\partial \Phi} = -\left(\frac{\partial D_0}{\partial \dot{\Phi}}\right) \tag{5.8}$$

i.e.

$$t_c^2\ddot{\varphi} + t_c\dot{\varphi}/Q + (\varphi-S)/A + \cos E \sin\varphi = 0 \tag{5.9}$$

where $\quad t_c = \sqrt{CL_J} \quad \text{and} \quad Q = R\sqrt{C/L_J} \tag{5.10}$

This equation of motion portrays the dynamics of the flux as a damped oscillating resonant circuit. By proper tuning of the Q factor the oscillation can smoothly be damped out as the excitation E going up to its full value of π.

5.2 Operational Principle of DCFP

The following qualitative illustration of the circuit operation will elicit the essential features of the circuit.

The location of the extremum potential in the range of $\pm 2\pi$ is given by,

$$\frac{\partial U_o}{\partial \varphi} = 0 \quad i.e. \quad \varphi - S + A\cos E \sin\varphi = 0 \tag{5.11}$$

If the second order derivative of U_o is greater (or smaller) than zero then there exists a minimum (or maximum) potential, else the potential has a transition point given by,

$$\frac{\partial^2 U_o}{\partial \varphi^2} = 0 \quad i.e. \quad 1 + A\cos E_c \cos\varphi_t = 0 \tag{5.12}$$

Given an input signal S and prior to the application of excitation E, the potential looks like Fig.5.2(a), with only an absolute minimum. When E is applied and is rising to a critical value E_c, the potential changes to Fig.5.2(b) with an absolute minimum and a turning point φ_t. When $E > E_c$ double-well potential emerges as shown in Fig.5.2(c). The maximum at φ_p is the potential barrier between the metastable minimum φ_m and the absolute minimum φ_a. The device is operated by clock excitation signal of amplitude π connected to E. When the clock signal rises up to its full value of π the output signal reaches the steady state and the amplification of the output signal is given by φ_a/S. The logic state is designated by the polarity of φ_a. On the other hand, if

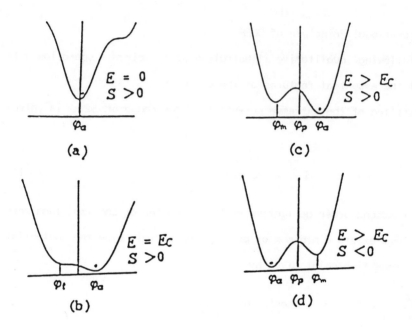

Potential States

Fig.5.2 (a) The potential state prior to apply clock excitation, (b) the potential state at transition state, (c) the potential state at transition state, (d) the potential state after transition state with inverted input seed signal.

inverse polarity input signal $-S$ is applied then $-\varphi_a$ is obtained as an inverse phase logic state as shown in Fig.5.2(d). The device characteristics emulate the conventional parametron device but operates on dc flux using a pair of Josephson junctions.

5.3 Static Characteristics of DCFP

The analysis of the device characteristics can be looked from another perspective, that is to analyse the φ vs S curve which is plotted based on (5.11). Since (5.11) gives the variation of the extremum points with respect to the change of E, thus the static behaviour and some of the qualitative dynamic features of the flux can be understood by analysing the plotting. The plotting also helps to define some of the useful parameters in the device as given in the following.

Fig.5.3 shows a plotting of φ vs S with curves A, B and C and straight lines SO and SL. SO is a perpendicular line to S-axis and the horizontal coordinate of this line represents the amount of input flux to the DCFP. Thus the line SO will be called source line hereinafter. The curves A, B and C are the plots of (5.11) with different E parameter values. The coordinate of intersection of the source line SO to the curve A, B or C represents a pair of $(input, output)_E$ where E denotes a given instance of clock excitation. For example curve A is plotted with $E=0.0$, and the vertical coordinate of intersection of the source line SO to the curve A represents the output flux at the instance of $E=0.0$ with input source given by the horizontal coordinate of intersection. Similarly curve B is a plot with $E=E_c$, where E_c is the critical E value. Thus source line SO tangents to curve C at one point and intersects it at another point. Curve C is a plot with $E=\pi$ and

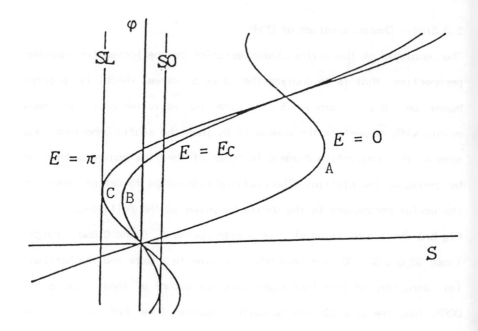

Fig.5.3 Plot of output signal verses input signal, where SL is a critical loading line and SO is the sourcing line.

with SL as the tangent line at the turning point of C. If the logic loading(which will be explained in more details later) is going up to or over this line the output flux would go into the wrong polarity, producing a wrong logic output. Thus the horizontal coordinate of this straight line denoted by S_L is defined as the critical logic loading. Also if the device parameter A is too large, the source line would also intersect the curve A at $\varphi > \pi$ thus giving rise to higher mode operation. The logic loading and the higher mode problems will be treated later.

5.4 Input and Output Correlations

In the following the relationship of the output signal φ_{out} to the input signal S will be derived by considering the problem of logic loading.

Fig.5.4(a) denotes a DCFP by a circle, and p input lines and n output lines are drawn on the left hand side and the right hand side of the circle respectively. The circuit operates on the majority logic principle. Assuming that k logic units consist of p input lines. Then in order to get the proper output amplification the output signal should be related to the input signal as follows,

$$S = \varphi_{out}/M \tag{5.13}$$

where $M=p+k$. The output signal φ_{out} is obtainable by substituting the above expression into (5.11) with $E=\pi$, that is

$$\varphi_{out}(1-\frac{1}{M}) - A\sin\varphi_{out} = 0. \tag{5.14}$$

Fig.5.4(b) shows a possible way to construct a simplest majority logic circuit with DCFPs coupling together. In this figure lead e is the

84

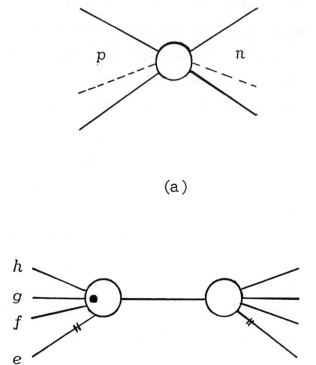

Fig.5.4 (a) A general DCFP logic circuit with p inputs and n outputs, (b) a three inputs majority logic circuit with booster circuit for logic signals distribution.

clock input, each circle represents a DCFP and a dot in the circle denotes that one majority logic unit is constructed by e, f, g and h leads connected to the DCFP. Thus according to (5.13) we get $S=\varphi_{out}/5$ with $M=5$.

5.5 Logic Loading and Higher Mode Hazards

The definition of A was given by (5.5) and the permissible range of A values can be obtained by analysing the logic loading hazard and the higher mode hazard.

Considering the logic loading condition in Fig.5.5(a), the H(high) logic state of the DCFP would be inverted to L(low) if the two H inputs turn off earlier than the L input under the condition that $S>-S_L$, where S_L was defined in the previous section and can be interpreted as the amount of net logic loading exert by all the leads connected to the DCFP at the steady state of $E=\pi$. The source line is shifted by the logic loading to the position of S_L, thus causing the potential to be distorted into a transition state as shown in Fig.5.5(c).

Similarly, the logic loading condition shown in Fig.5.5(b) could also cause the DCFP in the preceding stage to turn from H to L if $S>-S_L$. To ensure that logic loading may not result in wrong logic operation, we set $S_L \geq -2S$ with a safety factor of 2.

Thus if we denote the lower bound of A and its corresponding output by A_l and $\varphi_l (<\pi)$ respectively then they are obtainable by (5.11), (5.12) and (5.13) with $S_L=-2S$ and $E=\pi$ as follows,

$$\varphi_l (1+\frac{2}{M}) - \tan\varphi_l = 0 \qquad (5.15)$$

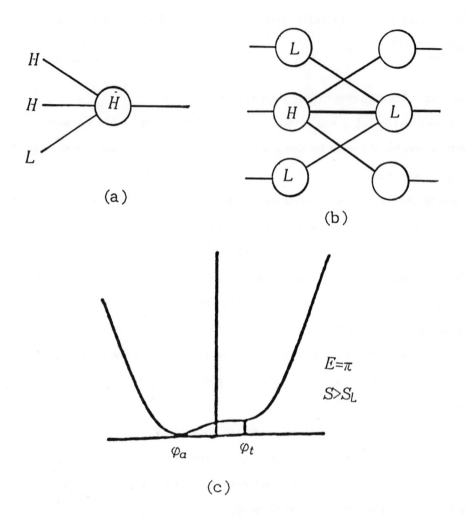

Fig.5.5 (a) and (b) are examples of circuit configurations which may have wrong logic operation if the magnitude of input signal is larger than the critical logic loading. Potential of critical logic loading is shown in (c).

$$A_l = \frac{1}{\cos\varphi_l} \tag{5.16}$$

The above expressions give the lower bound to the A value. On the other hand the A value also has an upper bound. With reference to Fig.5.3 it is clear that if A is too large the source line could also intersect the curve A (E=0.0) at $\varphi > \pi$, resulting in higher mode operation. If we denote the upper bound of A and its corresponding output by A_u and $\varphi_u (>\pi)$ respectively then A_u is a parameter which makes the source line with $S=\pi/M$ tangents to curve A at φ_u and that is the higher mode state at E=0.0. According to (5.11), (5.12) and (5.13) we obtain,

$$\varphi_u - \frac{\pi}{M} - \tan\varphi_u = 0 \tag{5.17}$$

$$A_u = -\frac{1}{\cos\varphi_u} \tag{5.18}$$

Thus the range of A is given by $A_l < A < A_u$.

5.6 Leakage Inductances in Transformer Coupling

DCFP is excited by clocking signal via a transformer coupling. Practically there is always some leakage inductance in the transformer coupling circuit as shown in Fig.5.6(a).

In the following we shall prove a useful lemma which states that:

The circuit configuration given by Fig.5.6(a) is equivalent to the circuit configuration of Fig.5.6(b) if the leakage inductance $L_{e1}=L_{e2}$.

Prove: The potential energy of Fig.5.6(a) is,

$$U_l = \frac{(\varphi_3 - S)^2}{2L^*} + \frac{(e-E)^2}{2L_{e0}} + \frac{\varphi_n^2 + \varphi_m^2}{2L_{e1}} - A(\cos\varphi_1 + \cos\varphi_2) \tag{5.19}$$

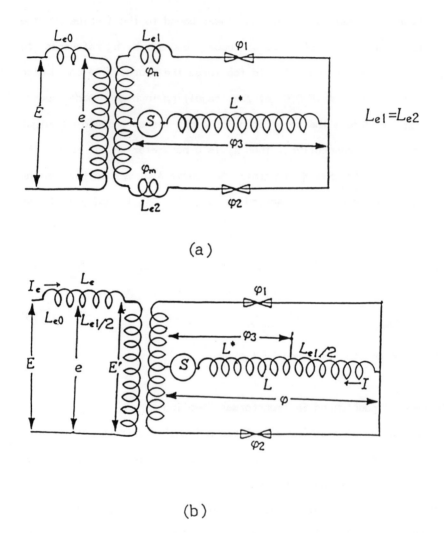

Fig.5.6 (a) DCFP with leakage inductance in the transformer coupling, (b) DFCP with equivalent leakage inductance in the transformer coupling.

if we define,

$$\varphi_2 + \varphi_1 = 2\varphi, \qquad \varphi_2 - \varphi_1 = 2E^{\cdot} \qquad (5.20)$$

and with reference to Fig.5.6(a) we get,

$$\varphi_1 = \varphi_m + \varphi_3 - e, \qquad \varphi_2 = \varphi_n + \varphi_3 + e \qquad (5.21)$$

according to (5.20) and (5.21) we get,

$$\varphi_n + \varphi_m = 2(\varphi - \varphi_3), \qquad \varphi_n - \varphi_m = 2(E^{\cdot} - e) \qquad (5.22)$$

thus

$$\varphi_n^2 + \varphi_m^2 = 2(\varphi_3 - \varphi)^2 + 2(E^{\cdot} - e)^2 \qquad (5.23)$$

Substituting this into U we get,

$$U_l = \left(\frac{(\varphi-\varphi_3)^2}{L_{e1}} + \frac{(\varphi_3-S)^2}{2L^*}\right) + \left(\frac{(E^{\cdot}-e)^2}{L_{e1}} + \frac{(e-E)^2}{2L_{e0}}\right) - 2A\cos\varphi\cos E^{\cdot} \qquad (5.24)$$

$$= \frac{I^2}{2}\left(\frac{L_{e1}}{2}+L^*\right) + \frac{I_e^2}{2}\left(\frac{L_{e1}}{2}+L_{e0}\right) - 2A\cos\varphi\cos E^{\cdot} \qquad (5.25)$$

$$= \frac{(\varphi-S)^2}{2L} + \frac{(E-E^{\cdot})^2}{2L_e} - 2A\cos\varphi\cos E^{\cdot} \qquad (5.26)$$

where $\quad L = \dfrac{L_{e1}}{2} + L^* \qquad L_e = \dfrac{L_{e1}}{2} + L_{e0} \qquad (5.27)$

and I and I_e are the current flows shown in Fig.5.6(b). The above final result is exactly equal to the potential energy of circuit shown in Fig.5.6(b) given in terms of new variables. The kinetic energy and the heat dissipating function of these two circuits are obviously equivalent in accordance with the definition of the new variables. Thus the two circuits are equivalent to each other. QED

5.7 DCFP with Leakage Inductances

In the last section we have found a simplified equivalent DCFP device with leakage inductances. In the following the simplified circuit of

Fig.5.6(b) will be used to derive the equations of motion of the flux for a non-ideal transformer coupling. In Fig.5.6(b) if we define the leakage inductance $L_e = BL_J/2$, where B is a parameter, then the potential energy, the kinetic energy and the heat dissipating function can be written respectively as,

$$U_l = E_J((E-E')^2/B + (\varphi-S)^2/A - 2(\cos E' \cos\varphi)) \tag{5.28}$$

$$K_l = \frac{C}{2}((\dot\Phi - \dot\Phi_E')^2 + (\dot\Phi + \dot\Phi_E')^2) \tag{5.29}$$

$$D_l = \frac{(\dot\Phi - \dot\Phi_E')^2 + (\dot\Phi + \dot\Phi_E')^2}{2R} \tag{5.30}$$

Thus the equations of motion of the flux are,

$$t_c^2 \ddot E' + t_c \dot E'/Q + (E'-E)/B + \sin E' \cos\varphi = 0 \tag{5.31}$$

$$t_c^2 \ddot\varphi + t_c \dot\varphi/Q + (\varphi-S)/A + \cos E' \sin\varphi = 0 \tag{5.32}$$

5.8 The Supercurrent Noise

So far in our analysis we have omitted the I_m deviation to be called δI_m noise or supercurrent noise. The δI_m noise is attributed to the imperfection junction fabrication or some changes of physical conditions in the junction. Assuming that I_m is deviated by δ from its designated value, then the worst deviated potential (excluding leakage inductance) would be,

$$U_s = \frac{(\Phi-\Phi_S)^2}{2L} - (\frac{I_m \Phi_o (1\pm\delta)}{2\pi}\cos\varphi_1 + \frac{I_m \Phi_o (1\mp\delta)}{2\pi}\cos\varphi_2) \tag{5.33}$$

or $U_s = E_J((\varphi-S)^2/A - 2(\cos E' \cos\varphi \mp \delta \sin E' \sin\varphi))$ (5.34)

Thus the equation of motion (5.9) becomes,

$$t_c^2 \ddot\varphi + t_c \dot\varphi/Q + (\varphi-S)/A + \cos E' \sin\varphi \pm \delta \sin E' \cos\varphi = 0 \tag{5.35}$$

The potential given by (5.33) is different from the potential (5.7) by the term with δ. If δ is too large the device characteristic given by (5.7) as discussed previously will be lost. Thus the circuit can not work as a good logic device. We define δ_c as the critical δ which will degrade the circuit potential to a single-minimum-well potential (for φ in the range of $\pm 2\pi$) at $E'=E_c'$. By considering the aforesaid conditions, we set the first, second and third order derivatives of U_s simultaneously equal to zero at $E'=E_c'$ as follows,

$$\varphi - S + A(\cos E_c' \sin\varphi + \delta \sin E_c' \cos\varphi) = 0 \qquad (5.36)$$

$$1 + A(\cos E_c' \sin\varphi + \delta \sin E_c' \cos\varphi) = 0 \qquad (5.37)$$

$$A(\cos E_c' \sin\varphi + \delta \sin E_c' \cos\varphi) = 0 \qquad (5.38)$$

From the above three equations we obtain,

$$\varphi_c = S, \qquad E_c' = \pi - \cos^{-1}\frac{\cos S}{A},$$

$$\delta_c = \pm \frac{\sin S}{\sqrt{(A-\cos S)(A+\cos S)}} \qquad (5.39)$$

A good device should have δ smaller than δ_c. Thus for a good device, E_c', φ_t and φ_a, by their definitions, are obtainable from the following equations which are derived from (5.36) to (5.38) as follows,

$$(1-\delta^2)\big((\varphi_t-S)\sin\varphi_t+\cos\varphi_t\big)^2 - (\varphi_t-S)^2 + A^2\delta^2 - 1 = 0 \qquad (5.40)$$

$$\varphi_a - S + A(\cos E_c' \sin\varphi_a + \delta \sin E_c' \cos\varphi_a) = 0 \qquad (5.41)$$

$$\cos^2 E_c' = \frac{(\varphi_t-S)^2 - A^2\delta^2 + 1}{A^2(1-\delta^2)} \qquad (5.42)$$

Now if we include the leakage inductance into the device it can be shown by refering to the last section that the equations of motion become,

$$t_c^2 \ddot{E}' + t_c \dot{E}'/Q + (E'-E)/B + \sin E' \cos\varphi \pm \delta \cos E' \sin\varphi = 0 \qquad (5.43)$$
$$t_c^2 \ddot{\varphi} + t_c \dot{\varphi}/Q + (\varphi-S)/A + \cos E' \sin\varphi \pm \delta \sin E' \cos\varphi = 0 \qquad (5.44)$$

5.9 Reduction of Supercurrent Noise by Split Excitations

In the previous section we pointed out that δIm noise could impair the logic function of the device. Thus it is essential that either technology of fabrication can reduce the δIm noise or some modification to the DCFP circuit can improve the circuit tolerance to the δIm noise. In the fabrication technology we can either use laser beam to trim the Josephson junction or improve the process technology. On the other hand it is possible to modify slightly on the DCFP circuit so as to improve the circuit tolerance to the δIm noise as discussed in the following.

Fig.5.7(a) is a DCFP device with split excitation driven by a double clocking excitation E_1 and E_2 with waveform given in Fig.5.7(b). Noting that Fig.5.7(a) is equivalent to Fig.5.1(a) if E_2 is taken away. The potential energy of this device is,

$$U_d = E_L \left(\frac{(\varphi-S)^2}{2} - A_1 (\cos E_1 \cos\varphi \pm \delta_1 \sin E_1 \sin\varphi) - A_2 (\cos E_2 \cos\varphi$$
$$\pm \delta_2 \sin E_2 \sin\varphi)) \qquad (5.45)$$

where $E_L = \frac{\Phi_0^2}{4\pi^2 L^2}$, $L = \frac{A_1}{2} L_{J1} = \frac{A_2}{2} L_{J2}$ \qquad (5.46)

In this device the second excitation clock signal E_2 is maintained at zero until the first clock reaching the value of π, thereafter it will start to excite from zero to π. Thus it is called DCFP with split excitation. Similar to the previous reasoning, by setting $E_2=0.0$ and the first to third derivatives of U_d equal to zero we obtain

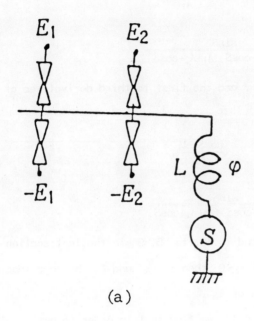

(a)

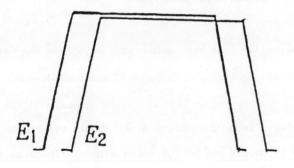

(b)

Fig.5.7 (a) Split excitation DCFP circuit, (b) split excitation clock signal to drive the DCFP.

$$\varphi = S, \quad \cos E_{1c} = -\frac{\cos S + A_2}{A_1},$$

$$\delta_{1c} = \pm \frac{\sin S}{\sqrt{(A_1 - A_2 - \cos S)(A_1 + A_2 + \cos S)}} \quad (5.47)$$

Also by setting $E_1 = \pi$ and the first to third derivatives of U_d equal to zero we obtain,

$$\varphi = S, \quad \cos E_{2c} = -\frac{\cos S - A_1}{A_2},$$

$$\delta_{2c} = \pm \frac{\sin S}{\sqrt{(A_2 - A_1 + \cos S)(A_2 + A_1 + \cos S)}}. \quad (5.48)$$

Comparing (5.47) and (5.48) to (5.39) in the last section it is clear that if $A = A_1$ and $A_1 > A_2 \neq 0$ then $\delta_{1c} > \delta_c$ and δ_{2c} becomes imaginary which means non-existence of δ_{2c}.

From (5.47) and (5.48) we find that in order to get a larger δ_{1c} the following relationship should be established,

$$A_1 - A_2 = \cos S. \quad (5.49)$$

For practical application S is very small thus we obtain the following approximation,

$$A_1 - A_2 \doteq 1. \quad (5.50)$$

Considering the logic loading problem, $A_1 + A_2$ should not be too small. On the other hand it should not be too large either, otherwise the δ_{1c} and δ_{2c} would be too small as implied by (5.47) and (5.48). Though we have figured out the range of desirable A_1 and A_2 values, the optimal values of A_1 and A_2 shall be worked out by dynamic simulation.

δIm noise can be classified into two types, one is the initial variation of Im due to imperfection of fabrication process, the other is due to heat cycling and aging of material after the junctions have been made. It is worth noting that the split excitation method can cover the defects originated from both of these causes, whereas improvement of trimming or fabrication process can only cover the former. However the choice of simple DCFP or split excitation DCFP is up to the user's requirements and the kind of technology involved. Of course simple DCFP is preferred if the material of the junctions and the technology of fabrication can somehow effectively reduce the δIm noise.

5.10 Thermal Noise at Transition State

When a seed signal S is given to the device, then clock signal is applied to amplify the seed signal to reach the maximum output signal at $E=\pi$. As clock excitation is applied the time varying potential changes from single potential well to double potential well. If the clock signal is rising slowly then the flux 'particle' should be located around φ_a while the potential is changing. However thermal energy may bring the flux 'particle' over the barrier to the metastable location during this changing. The most critical thermal noise is at $E=E_c$, and the probability of it is given by the Boltzmann factor as follows,

$$P = \frac{1}{2}e^{-V/kT} = \frac{1}{2}e^{-(U(\varphi_t)-U(\varphi_a))/kT} \tag{5.51}$$

where the 1/2 coefficient is necessary for reason that after the thermal excitation the flux 'particle' has approximately equal probability to decay into either one of the logic states. Numerical values of φ_t and φ_a are obtainable from equations (5.40) to (5.42) and the probability of thermal noise can thus be obtained.

If we assume that the δIm noise is negligible small then φ_t and φ_a can be obtained by substituting (5.12) into (5.11) as follows,

$$\varphi_t - \tan\varphi_t = S \qquad (5.52)$$

and

$$\varphi_a - \frac{\sin\varphi_a}{\cos\varphi_t} = S \qquad (5.53)$$

Defining $G = (3S)^{1/3}$ and using computer algebra system<Moritsugu84> we find that

$$\varphi_t = -G(1 - \frac{2G^2}{15} + \frac{3G^4}{175} - \frac{2G^6}{1575} - \frac{16G^8}{202125}) \qquad (5.54)$$

$$\varphi_a = G(2 - \frac{G^2}{6} + \frac{3G^4}{200} - \frac{11G^6}{18000} - \frac{1303G^8}{11760000}) \qquad (5.55)$$

$$P = \frac{1}{2}\exp\{-\frac{E_J G^4}{AkT}(\frac{9}{8} - \frac{9G^2}{80} + \frac{243G^4}{22400} - \frac{101G^6}{224000})\} \qquad (5.56)$$

5.11 Quantum Tunnelling and Thermal Noise

Now we shall consider the problem of quantum tunnelling at the steady state. At the steady state of $E=\pi$ the logic state is designated by the absolute minimum potential flux denoted by φ_a. Since the operation of DCFP is bilateral, there is always a feedback of signal from one stage to the preceding stage to provide a logic loading to the preceding stage as being discussed. Considerable amount of feedback signal would push up the absolute minimum potential to become a metastable potential such that φ_a no longer stands for the location of absolute minimum potential as shown in Fig.5.8. Under such a circumstance quantum tunnelling and thermal excitation of the flux may occur, resulting in wrong logic signal output. In the following we shall

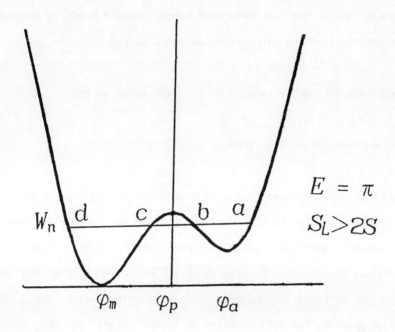

Fig.5.8 Potential of the DCFP with logic loading so that flux 'particle' is located at the metastable minimum (which should be the absolute minimum if there is no logic loading) thus tunnelling could occur.

treat the thermal-quantum tunnelling problem.

The phase differences in the Josephson junctions are given by<Josephson62>,

$$[n_1, \varphi_1] = 1 \quad \text{and} \quad [n_2, \varphi_2] = 1 \tag{5.57}$$

where n_1 and n_2 are the numbers of Cooper pairs tunnelling through the two junctions of the circuit. Since $V=q/C=en/C=\Phi_o\dot\varphi/2\pi$, we get,

$$k[\dot\varphi_1, \varphi_1] = 1 \quad \text{and} \quad [\dot\varphi_1, \varphi_2] = 1 \tag{5.58}$$

where $k=C\Phi_o/2\pi$. Substitute (5.6) into the above we get,

$$k[\dot E, E] + k[\dot\varphi, \varphi] = 1 \tag{5.59}$$

In the steady state $\dot E=0$, thus,

$$k[\dot\varphi, \varphi] = 1 \tag{5.60}$$

Expression (5.60) shows that φ is quantized. Noting that at $E=\pi$ term δ does not come into the the Hamiltonian thus we can write down the Schrodinger equation using the Hamiltonian (5.3). This is a one dimension quantization problem which can be solved by the WKB method which can be found in many text books of quantum theory. The quantized energy level W_n for $W_n \leq W_b$, where W_b is the height of the potential barrier, is given by,

$$J_n = 2\int_a^b \sqrt{4C(W_n-U_o)}\,d\Phi = (n+\frac{1}{2})h \qquad (n=0,1\ldots) \tag{5.61}$$

i.e.

$$4E_J t_c h^{-1}\int_a^b \sqrt{W_n-(\varphi-S)^2/A+2\cos E\cos\varphi}\,d\varphi = (n+\frac{1}{2}) \tag{5.62}$$

The transmitivity by tunnelling of quantum energy states is given by,

$$\gamma_n = exp\{-2\hbar^{-1}\int_b^c \sqrt{4C(U_o-W_n)}\,d\Phi\} \tag{5.63}$$

$$=exp\{\frac{-4E_Jt_c}{\hbar}\int_b^c \sqrt{(\varphi-S)^2/A-2cosEcos\varphi-W_n'}\,d\varphi\} \qquad (5.64)$$

where $W_n=E_JW_n'$, b to c is barrier thickness and a to b is the width of the potential well at energy state W_n.

The period of oscillation is,

$$\tau_n = \frac{\partial J_n}{\partial W} = \int_a^b \frac{2t_c d\varphi}{\sqrt{W_n'-(\varphi-s)^2/A+2cosEcos\varphi}} \qquad (5.65)$$

The rate of decay and the life time of the quantum flux at energy state W_n are respectively given by,

$$\rho_n = \gamma_n/\tau_n \quad \text{and} \quad \lambda_n=1/\rho_n \qquad (5.66)$$

Taking into account both the thermal excitation and quantum tunnelling, the average rate of decay and the average life time of the quantum flux are given approximately by,

$$\bar{\rho} = \frac{\sum_{W_n \leq W_b}\rho_n e^{-W_n/kT}+ \nu_o e^{-W_b/kT}/(1-e^{-h\nu_o/kT})}{\sum_{W_n \leq W_b} e^{-W_n/kT} + e^{-W_b/kT}/(1-e^{-h\nu_o/kT})} \qquad (5.67)$$

$$\bar{\lambda} = \frac{\sum_{W_n \leq W_b}\lambda_n e^{-W_n/kT}+\tau_o e^{-W_b/kT}/(1-e^{-h\nu_o/kT})}{\sum_{W_n \leq W_b} e^{-W_n/kT} + e^{-W_b/kT}/(1-e^{-h\nu_o/kT})} \qquad (5.68)$$

where $\tau_o=1/\nu_o=2\pi t_c$.

5.12 Simulations and Numerical Analysis

To simulate the logic operation of the DCFP devices it is necessary to have a clock. In the following simulation we assume a dc biased sinusoidal clock signal to be used. That is

$$E = \frac{\pi}{2}(C_1 - C_2 \cos\frac{2\pi t}{T_p}), \qquad (5.69)$$

where T_p is the period of the clock. If $C_1=1.0$, $C_2=1.0$ and $T_p=100.0ps$ then it will be called the standard clock as shown in Fig.5.9(a).
According to the definition of power dissipation function given by (5.2) the average power dissipation is obtainable as follows,

$$P_V = \frac{2}{T_p}\int_0^{T_p} D dt = \frac{1}{RT_p}\int_0^{T_p}((\dot{\Phi}-\dot{\Phi}_E)^2 + (\dot{\Phi}+\dot{\Phi}_E)^2) dt \qquad (5.70)$$

where $\dot{\Phi}$ and $\dot{E}$ are obtainable numerically from (5.43) and (5.44). Computer programs were developed to simulate the circuit operation and to analyse the device parameters as given in the following.

(i) If we assume that DCFP is used to make a simplest majority logic circuit then according to the definition of M given in (5.13) and the discussion therein we get $M=5$. By (5.16) and (5.18) we get $A_l=1.61$ and $A_u=3.96$. If we choose $A=2.25$ which is within the range of A_l and A_u then the output is obtainable by (5.14) to be $\varphi_{out}=2.23$, the input by (5.13) to be $S=0.45$ and the critical δI_m noise by (5.39) to be $\delta_c=0.21$.

(ii) A resonant circuit with circuit dynamic given by (5.43) and (5.44) can have a smooth output signal if Q is properly chosen. In the following $Q=0.5$ will be used to carry out the simulation. A typical Josephson tunnel junction has $I_m R = 1.0mV$, thus if we assume $I_m=100uA$ then using (5.5) and (5.10) we obtain the following device parameters as $L_J=3.18pH, C=0.008pF$ and $R=10.0\Omega$.

(iii) The probability of thermal noise is given by (5.51), where φ_t

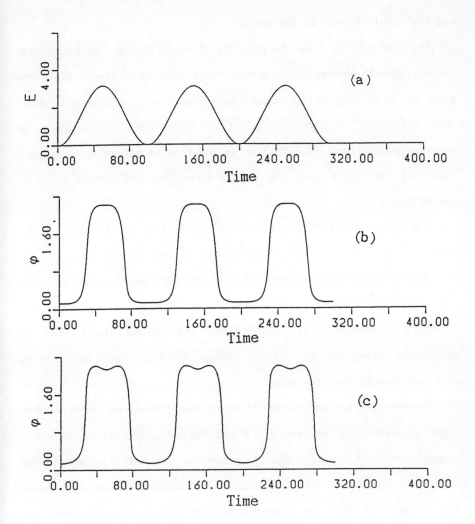

Fig.5.9 (a) a standard dc bias sinusoidal clock signal used for simulation, (b) simulated output of a ideal DCFP, (c) simulated output of a DCFP with 40% of leakage inductance and 10% of δIm noise.

and φ_a are obtainable from (5.40) to (5.42). If $T=4.2°K$ then numerical calculation shows that $P=10^{-185}$, which is very small thus thermal noise has negligible effect on the device.

(iv) The average life time and the rate of decay of the flux by quantum tunnelling with thermal noise due to logic loading at steady state are given by (5.67) and (5.68) respectively. Numerical calculation shows that $\bar{\rho}=1.14\times10^{-73}/s$ and $\bar{\lambda}=5.29\times10^{74}s$. Thus the rate of decay is negligibly small and the life time of the flux is extremely long, thus there is no worry about the quantum tunnelling and thermal noise at steady state.

(v) Simulations of DCFP operation based on (5.43), (5.44) and (5.69) with $C_1=1.0$, $C_2=1.0$, $\delta=0.0$ and $B=0.0$, and with $C_1=1.3$, $C_2=1.3$, $\delta=\delta_c/2=0.10$ and $B=0.40$ are plotted in Fig.5.9(b) and Fig.5.9(c). The power dissipations were calculated based on (5.70) to be $P_w=1130pW$ and $P_w=2700pW$ respectively. The simulations demonstrate that within a reasonable deviation of circuit parameters the circuit can operate well and consume very low power.

(vi) To simulate the operation of split excitation we need double clock of waveform as shown in Fig.5.7(b). Fig.5.10(a) is the clock signal with the simulation result given in Fig.5.10(b), where the simulation was done with $A_1=2.0$, $A_2=1.0$ and $\delta=1.00$. From the simulation we conclude that the circuit almost immunes to the δIm noise.

(vii) A circuit simulation code generator was developed to simulate the circuit operation for multiple DCFPs interconnected via some delay lines. The generator and its applications for DCFPs circuits simulation will be studied in the next chapter.

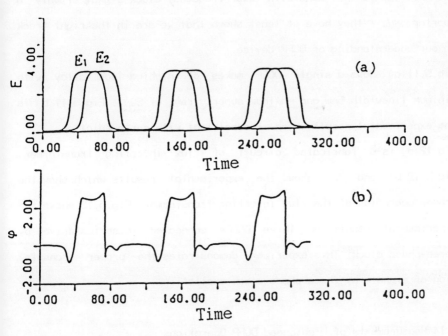

Fig.5.10 (a) a clock signal to be used for simulation on split excitation DCFP, (b) simulated output of split excitation DCFP. The result shows that the device works even up to 100% of δIm noise.

5.13 Experiments on DCFP Circuits

It is exciting to report that Hitachi Central Research Laboratory has recently carried out experiments<Harada85> to confirm that DCFP can actually operate as predicted by our analysis and simulations. Though the experiments were done with lower frequency clock signal of only a quarter MHz, they have at least shown that we are in the right track in our understanding of DCFP device.

Fig.5.11(a) shows a single DCFP makes of niobium-lead alloy with minimum linewidth $5\mu m$ and maximum supercurrent of $50\mu A$. Fig.5.11(b) is the experimental result which is consistent with our prediction.

Fig.12(a) is a fabricated circuit of flux inverting transformer. Fig.5.12(b) and (c) show the experimental results which show the proper operation of the flux inverting transformer. Fig.5.13 shows the experimental result on three DCFPs connected in series driven by three-phase clock. This experiment demonstrates the proper transmission of signals by three phase clock excitation.

5.14 Measurements of High Speed DCFP Operations

According to simulation that DCFP is supposed to work at 10GHz. However experiment to measure at $10GHz$ is not easy. Following methods to observe the proper operation of DCFP at high speed called $ring\ counter\ methods$ are proposed.

Consider a ring counter consisting of three DCFPs with one negation

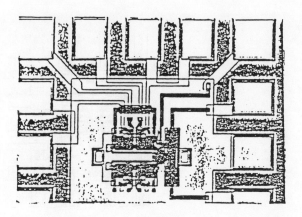

(a)

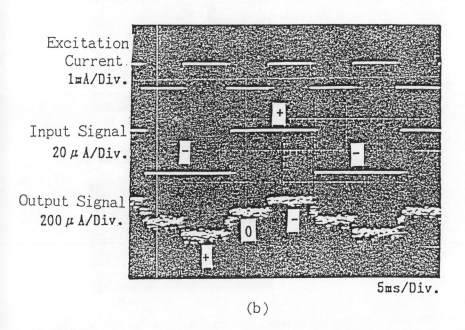

(b)

Fig.5.11 (a) a fabricated DCFP device, (b) the experimental result on single DCFP operation.

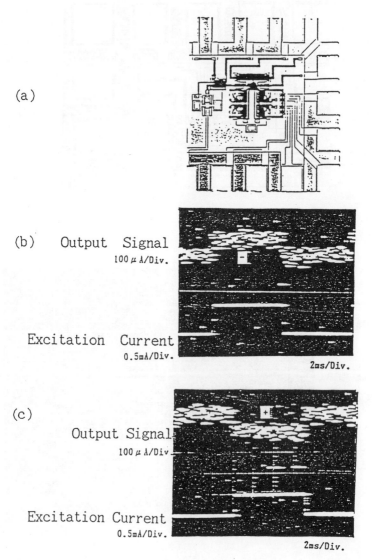

Fig.5.12 (a) a fabricated flux inverting transformer for NOT logic (b) the output result with a positive input, (c) the output result with a negative input.

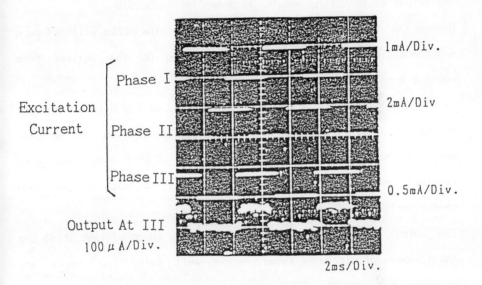

Fig.5.13 Experiment on three DCFP connected in series driven by 3-phase clock, the result of the experiment indicates that the 3-phase clock is successfully driving forward the flux signal in the DCFPs.

circuit as shown in Fig.5.14(a). Let the clock frequency be F, then the output at the last DCFP as shown in Fig.5.14(b) is alternating with subharmonic frequency $F/2$. Thus the circuit will act as a 1/2-subharmonic generator independence of the initial conditions. A ring counter consisting of six DCFPs as shown in Fig.5.15(a) will act as a 1/4-subharmonic generator independence of the initial conditions. The output of this ring counter is shown in Fig.5.15(b).

However a ring counter like Fig.5.16(a) with nine DCFPs will act as a subharmonic generator of $F/2$ or $F/6$ depending on the initial free running condition as shown in Fig.5.16(b).

It would be preferably to use the ring counter of Fig.5.14 or Fig.5.15 as the output of it does not depend on the initial value. The measurement of subharmonic generated at the output of DCFP with a microwave receiver tuned at the subharmonic frequency will be much easier than measuring the waveform.

The output power of a DCFP output is in the range of $1nW$, while the input levels of microwave receivers are well below $1pW$. Hence the coupling between the DCFP and the receiver may be rather loose. It would be worth noting that the outputs of a $faithful$ subharmonic generator are phase locked.

Hence if phase locking of the ring counter as of Fig.5.15 with 6 DCFPs at 10GHz be confirmed over a day ($10^5 sec$), then this will imply the correctness of 6×10^{15} DCFP operations.

In order to confirm the phase locking another 1/4-subharmonic generator is needed to provide the phase reference. Two stages of varactor diode parametrons each acting as a 1/2-subharmonic generator may be used for this purpose at the room temperature.

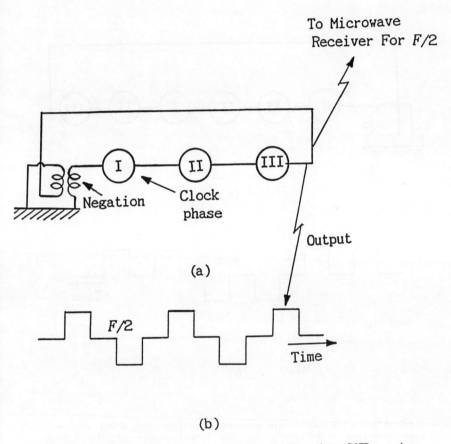

Fig.5.14 (a) a 1/2-subharmonic generator made from three DCFPs and a flux inverting transformer, (b) the output signal of the 1/2-subharmonic generator.

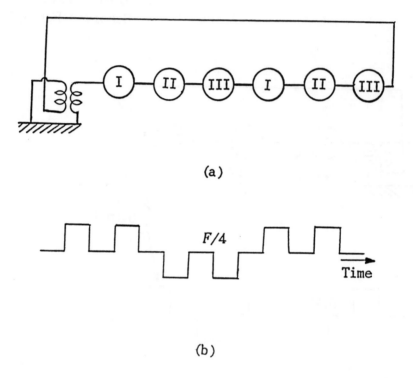

Fig.5.15 (a) a 1/4-subharmonic generator made from six DCFPs and a flux inverting transformer, (b) the output signal of the 1/4-subharmonic generator.

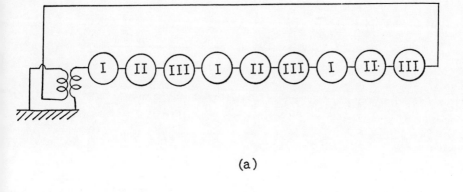

(a)

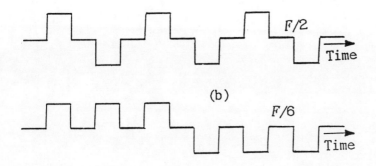

Fig.5.16 (a) a 1/2-1/6-subharmonic generator made from nine DCFPs and a flux transformer, (b) two possible output signal from the 1/2-1/6-subharmonic generator.

Chapter 6
CIRCUIT SIMULATION CODE GENERATOR

In the last chapter we have studied the basic operational principle of DCFP. Also we have done simulation on a single unit of DCFP operation. However utilizing DCFPs to realize more complicated logic circuit for computer, it is necessary to connect DCFPs together to make various logic circuits. It is not so difficult to write a program for simulating the operation of a single DCFP, however when more DCFPs are connected together making more complicated system then chances to make mistakes in programming are higher let alone the longer time required to write such programs. Also there are all kinds of circuits, which are constructed by combining DCFPs in various ways, need to be simulated. If for every circuit we have to write a program for simulation then the difficulties will be multiplied. Owing to all these reasons, a good software tool is indispensable for design simulations. In this chapter we shall introduce our idea of how to develop a circuit simulation system which is easy to use and well suited for our design purpose. The system is called Circuit Simulation Code Generator (CSCG).

6.1 Design Motivations and Characteristics of CSCG

There are a number of circuit simulation systems available in the market. Among all these systems, some use table driven method, others use compiled code method. Table driven method is analogue to program interpretation and compiled code method to program compilation. CSCG is designed based on the latter method, which is generally able to execute faster.

One of the unique characteristics of CSCG is that circuit dynamics is formulated by Hamiltonian approach. To our knowledge all the conventional circuit simulation systems use Kirchoff's current and voltage laws and device characteristics as a basic foundation to formulate circuit equations together with nodal analysis method, sparse tableau method, modified nodal analysis method etc. Another characteristic of our system is that the system is designed based on computer algebra system(we use REDUCE for system design). So far to our knowledge there is only a circuit simulation system named NETFORM is designed based on computer algebra system(also using REDUCE system). Notwithstanding, NETFORM is still based on the conventional circuit analysis method. Moreover, NETFORM does not have a facility to produce a complete program for analysis, and the user has to write a part of the program in order to analyse a circuit.

The motivations for us to design CSCG are,

- to provide a new alternative for circuit simulation system design,
- to provide a simulation system which is in parallel to mechanical approach, namely the Hamiltonian and Lagrangian approaches for Josephson circuit study.
- to provide a simple and economical methodology for user to construct a circuit simulation system based on some simple but elegant considerations.

The first point needs no elaboration. The second point is essential since in the DCFP study we take a mechanical view of the flux thus mental vision of the the flux behavior is possible. A circuit simulation system extended from this consideration would help us to extend a consistent view to a more complicated circuit system using DCFPs as the basic device. The last point is reflected in the fact that the basic part of this system was able to be designed within a week of working time due to the simple and elegant algorithms derivable from the idea of this approach. Thus this approach provides an economical means to obtain a circuit simulation system.

6.2 Principles and Criterions for CSCG Design

In the previous discussion of Lagrangian and Hamiltonian approaches to circuit dynamic theory, we note that the energy of a circuit system can be seperated into three basic energy localizations in individual physical elements namely capacitor, inductor and resistor. Owing to this characteristics, any graphical diagram of connections is also a pictorial schematic of the physical arrangement.

These basic features prompted us to design a circuit simulation system based on the Hamiltonian approach, since a Hamiltonian represents the energy of a system and if we add up the energy of all the components we can easily obtain the total Hamiltonian of the system. Besides, the equations of motion derived from the Hamiltonian is first order differentiate equations, it can straightforwardly be solved by some numerical analysis method such as the Runge Kutta method.

When designing the CSCG, we kept in mind that the system should confirm to some of the basic criterions in software engineering, that is user friendly and maintenanable with proper documentation. We found that it is surprisingly simple to realize these criterions based on the computer algebra and Hamiltonian approach. More specifically it is easy to implement the following ideas into the system:

- easy specification input by users to simulate a circuit,
- easy extension of the system if users need to incorporate new functions into the system,

- comprehensive graphical presentations of output,
- comprehensive and compact documentation.

We shall elaborate how all these can easily be implemented in the Hamiltonian approach using computer algebra to manipulate the algorithms

Since the energy of a circuit system is stored in the basic circuit elements, and the total energy of the system is given by the Hamiltonian. We can strategically decompose a circuit system into sub-Hamiltonian systems which are being used as the basic components for the simulation system design. For example a DCFP is a sub-Hamiltonian of the total system consisting of two Josephson junction and a flux transformer. A delay line consists of RCL coupling components of several stages. Using this as the basic units, computer algebra can be used to formulate for each of them and set into the system. As for a user, he only has to input specifications in order to use these basic components to compose his circuit configuration. Specifications consist of the names of the basic components and the interfacing parameters of the basic components in one to one correspondence with the physical arrangement of circuit components in a circuit or circuit diagram to be simulated.

Computer algebra algorithms in the CSCG will automatically manipulate the specifications input by the user to construct the Hamiltonian of the total circuit system and produces the formulas for circuit simulation based on some numerical approach for solving the equations. All the solutions are produced by the computer algebra system symbolically. Finally for efficiency in numerical simulation we generate the FORTRAN program from the symbolical solutions obtained by the algebra system. The computer algebra system we used is REDUCE, which has the facility to generate FORTRAN program as we shall see later.

Thus our first criterion for system design is already achieved by the abovementioned consideration since writing circuit specifications are just providing the names of the basic components and their interface parameters in accordance with the order of arrangement of the basic components in the circuit diagram. The second criterion also can easily be achieved since for a user to extend the system to include a new circuit component, he only has to write a sub-Hamiltonian of that component and code it in the computer algebra code following an example in the system. The third criterion of our design does not bear any relation to the approach we adopted, it is simply that there are many output data necessary to describe a simulation result; instead of tabulating the data, graphical presentations provide an immediate visual understanding of the dynamic behaviour of the circuit. Incidentally, the paper for output plotting can also contain the circuit specifications and the circuit parameters in a piece of paper serving as a compact document, meeting our last criterion. This document is comprehensible owing to the fact that the specifications are one to one corresponding to the physical arrangement of the circuit. Therefore, inversely, we shall be able to reproduce the circuit diagram based on this document, since all the simulation conditions of a circuit are captured in this document. Thus we have fulfilled all our design criterions and turned out a simple and elegant system.

6.3 Algebra Algorithms for CSCG

Using the full length of the program list to illustrate our system design would not only complicate the explanation but also eclipse the essential point of the system design. Therefore we shall come back to the simple harmonic oscillation system which we have studied in section 4.2 and write the algorithms for solving the Hamiltonian of this mechanical system by REDUCE to explain the underlying algebra algorithms being used to construct the system. The algebra algorithms based on REDUCE to solve the system dynamic of Fig.4.1 harmonic oscillator is given in List 1.

Line preceded with % is only a comment line. Lines 4-7 are the input to the system, they are potential energy, kinetic energy, dissipating function and Hamiltonian respectively. Lines 11-24 are the algebra algorithms for a procedure which produces a symbolical solution of a set of differential equations based on the Runge Kutta method. Wherein sub(...) is a REDUCE function which performs substitution to its last argument by all the preceding arguments. HH and TT are two parameters required in the Runge Kutta method; HH is the step of time interval for numerical analysis and TT is the total time of system evolution. Lines 28-29 are operations to obtain the equations of motion from the Hamiltonian which was given in Line 7. Wherein DF(..) is a system function of REDUCE to obtain partial differential of the first argument by the second argument. Line 30 calls for the Runge Kutta procedure which was given earlier. Lines 34-86 generates the FORTRAN program for numerical simulation.

List 2 is the FORTRAN program generated from LIST 1. Note that the algebraic Runge Kutta algorithms generate a symbolical Runge Kutta solution for substituting the labels in Lines 77-78 of List 1 to obtain the actual Runge Kutta solutions in Lines 41-46 of List 2.

List 1

File: DCFP.TEXT(SHM)   on Friday - 3 Aug 84 at   9.46.59   page : 1

```
 1: %
 2: %  INPUT
 3: %                                                                  ;
 4:   K := 1/(2*M)*P**2 $
 5:   U := K0/2*Q**2 $
 6:   D := B/2*QDOT $
 7:   H := K + U $
 8: %
 9: %  RUNGE-KUTTA METHOD
10: %                                                                  ;
11:  PROCEDURE RUNGEKUTTA(P1,P2,P,Q,TT);
12:   BEGIN
13:    SCALAR K11, K12, K21, K22, K31, K32, K41, K42;
14:       K11 := HH*P1;
15:       K12 := HH*P2;
16:       K21 := HH*SUB(TT=TT+HH/2, P=P+K11/2, Q=Q+K12/2, P1);
17:       K22 := HH*SUB(TT=TT+HH/2, P=P+K11/2, Q=Q+K12/2, P2);
18:       K31 := HH*SUB(TT=TT+HH/2, P=P+K21/2, Q=Q+K22/2, P1);
19:       K32 := HH*SUB(TT=TT+HH/2, P=P+K21/2, Q=Q+K22/2, P2);
20:       K41 := HH*SUB(TT=TT+HH  , P=P+K31  , Q=Q+K32  , P1);
21:       K41 := HH*SUB(TT=TT+HH  , P=P+K31  , Q=Q+K32  , P2);
22:       PN  := P + (K11 + 2*K21 + 2*K31 + K41)/6;
23:       QN  := Q + (K12 + 2*K22 + 2*K32 + K42)/6;
24:   END$
25: %    ;
26: %   HAMILTONIAN CALCULATION  ;
27: %                                                                  ;
28:   DIFQ := DF(H,P);
29:   DIFP := -DF(H,Q)- SUB( QDOT=P/M, DF(D,QDOT)) ;
30:   RUNGEKUTTA(DIFP, DIFQ, P, Q, TT);
31: %    ;
32: %  FORTRAN PROGRAM OUTPUT   ;
33: %                                                                  ;
34:   OFF ECHO$
35:   ON FORT$
36:   OUT F2;
37: WRITE "         PROGRAM RUNGE";
38: WRITE "*";
39: WRITE "*  INPUT";
40: WRITE "*";
41: WRITE "        IMPLICIT REAL(K,M)";
42: WRITE "        WRITE(6,*) 'INITIAL VALUE OF P'";
43: WRITE "        READ(5,*) P";
44: WRITE "        WRITE(6,*) ' P = ' P";
45: WRITE "        WRITE(6,*) 'INITIAL VALUE OF Q'";
46: WRITE "        READ(5,*) Q";
47: WRITE "        WRITE(6,*) ' Q = ' Q";
48: WRITE "        WRITE(6,*) 'VALUE OF M'";
49: WRITE "        READ(5,*) M";
50: WRITE "        WRITE(6,*) ' M = ' M";
51: WRITE "        WRITE(6,*) 'VALUE OF K0'";
52: WRITE "        READ(5,*) K0";
53: WRITE "        WRITE(6,*) ' K0 = ' K0";
54: WRITE "        WRITE(6,*) 'VALUE OF B'";
55: WRITE "        READ(5,*) B";
```

List 1

File: DCFP.TEXT(SHM)   on Friday 3 Aug 84 at 9.46.59   page : 2

```
56: WRITE "        WRITE(6,*) ' B = ' B";
57: WRITE "        WRITE(6,*) 'STEP SIZE OF T'";
58: WRITE "        READ(5,*) HH";
59: WRITE "        WRITE(6,*) ' STEP SIZE OF T = ' HH";
60: WRITE "        WRITE(6,*) 'FINAL VALUE OF T ?'";
61: WRITE "        READ(5,*) TP";
62: WRITE "        WRITE(6,*) ' FINAL VALUE OF T = ' TP";
63: WRITE "*";
64: WRITE "* INITIALIZATION";
65: WRITE "*";
66: WRITE "        TT = 0";
67: WRITE "        WRITE(9,*) ' H = ",H,"'"";
68: WRITE "        WRITE(9,*) ' D = ",D,"'"";
69: WRITE "        WRITE(9,901) C";
70: WRITE "    901 FORMAT(' C= ' E20.10)";
71: WRITE "        WRITE(9,910) TT,Q,P";
72: WRITE "    910 FORMAT(' ',3E20.10)";
73: WRITE "*";
74: WRITE "* LOOP";
75: WRITE "*";
76: WRITE "    100 CONTINUE";
77: WRITE "        PN=", PN;
78: WRITE "        Q=", QN;
79: WRITE "        P = PN";
80: WRITE "        TT= TT + HH";
81: WRITE "        WRITE(9,910) TT,Q,P";
82: WRITE "        IF( TT.LT.TF ) GO TO 100";
83: WRITE "        STOP";
84: WRITE "        END";
85: OFF;
86: OFF FORT$
87: SHUT F2;
88: ;END;
```

List 1 Algebraic algorithms for generating the a FORTRAN program to solve a harmonic oscilating system numerically.

List 2

File: DCFP.FORT77(SHM)   on Friday 3 Aug 84 at 9.47.17   page : 1

```
 1:          PROGRAM RUNGE
 2: *
 3: *   INPUT
 4: *
 5:          IMPLICIT REAL(K,M)
 6:          WRITE(6,*) 'INITIAL VALUE OF P'
 7:          READ(5,*) P
 8:          WRITE(6,*) ' P = ' P
 9:          WRITE(6,*) 'INITIAL VALUE OF Q'
10:          READ(5,*) Q
11:          WRITE(6,*) ' Q = ' Q
12:          WRITE(6,*) 'VALUE OF M'
13:          READ(5,*) M
14:          WRITE(6,*) ' M = ' M
15:          WRITE(6,*) 'VALUE OF K0'
16:          READ(5,*) K0
17:          WRITE(6,*) ' K0 = ' K0
18:          WRITE(6,*) 'VALUE OF B'
19:          READ(5,*) B
20:          WRITE(6,*) ' B = ' B
21:          WRITE(6,*) 'STEP SIZE OF T'
22:          READ(5,*) HH
23:          WRITE(6,*) ' STEP SIZE OF T = ' HH
24:          WRITE(6,*) 'FINAL VALUE OF T ?'
25:          READ(5,*) TP
26:          WRITE(6,*) ' FINAL VALUE OF T = ' TP
27: *
28: *   INITIALIZATION
29: *
30:          TT = 0
31:          WRITE(9,*) ' H = (M*Q**2*K0+P**2)/(2.*M)'
32:          WRITE(9,*) ' D = (B*QDOT)/2.'
33:          WRITE(9,901) C
34:      901 FORMAT(' C= ' E20.10)
35:          WRITE(9,910) TT,Q,P
36:      910 FORMAT(' ',3E20.10)
37: *
38: *   LOOP
39: *
40:      100 CONTINUE
41:          PN=(-20.*B*M**2*HH+2.*B*M*HH**3*K0-4.*B*M*HH**2+B*HH**4*
42:         . K0+48.*M**2*P-40.*M**2*Q*HH*K0-16.*M*P*HH**2*K0+8.*M*P*HH
43:         . +4.*M*Q*HH**3*K0**2-8.*M*Q*HH**2*K0-4.*P*HH**3*K0+2.*Q*HH
44:         . **4*K0**2)/(48.*M**2)
45:          Q=(-2.*B*M*HH**2+12.*M**2*Q+10.*M*P*HH-4.*M*Q*HH**2*K0-P*
46:         . HH**3*K0)/(12.*M**2)
47:          P = PN
48:          TT= TT + HH
49:          WRITE(9,910) TT,Q,P
50:          IF( TT.LT.TF ) GO TO 100
51:          STOP
52:          END
```

List 2 FORTRAN program generated from LIST 1.

The above example explains the algorithms underlying the system design. However CSCG is designed according to the criterion that when a user uses the system for circuit simulation he does not have to know about all these algorithms. That is the above algorithms are transparent to the user. In the next section we shall explain how we implement the system such that users only have to input specifications in order to use the system.

6.4 Implementation of Modular Design and Specifications

In the last section we have explained the basic algorithms underlying the system design. We also note that users do not have to write such algorithms but only to input the specifications. The basic idea of the whole system operation can be summarized as follows: When a user writes the specifications of a circuit and executes the CSCG then a FORTRAN code for simulation of that circuit is generated. Now if he executes the FORTRAN code it would prompt him to input the numerical values of the circuit parameters. After input of all the parameter values the FORTRAN code will produce an output which consists of the simulation results in graphical presentations with flux verses time axis. The specifications and the input parameter values are also produced in the same piece of plotting.

We shall use a typical example to explain how the specifications look like and how internally they are processed by the system. In a way the implementation and structure of the system will be illustrated by this example.

Fig.6.1 shows a circuit diagram of five DCFPs interconnected via some delay lines. This is in fact the majority logic circuit. The principle

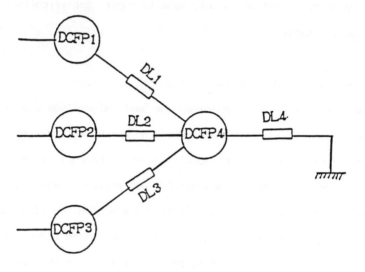

Fig.6.1 A DCFP majority logic circuit with three inputs and connected by delay lines.

is the same as the conventional parametron which we have explained in the earlier chapter. If we shall perform a simulation on this circuit starting from the basic formulation, then depending on the number of stages of delay lines being considered, we shall probably have to write the algorithms for solving more than 20 simultaneous first order differential equations. To use the CSCG for such a simulation the user only has to input essentially the following:

    DCFP(1,CK1);

    DCFP(2,ck1);

    DCFP(3,CK1);

    DL(1,X1,X4);

    DL(2,X2,X4);

    DL(3,X3,X4);

    DCFP(4,CK2);

    DL(4,X4,0);

On the above specifications we have not included the specifications of input and output controlled statements since they will be explained later. On the above, DCFP(p,CKn) is the specification of DCFP; the first argument is to designate the pth DCFP number and the second argument is to specify the phase of clock being used to drive the DCFP. DL(i,Xj,Xk) is the specification of a delay line; the first argument is an integer designating the ith delay line, and the second and the third arguments are interfacing flux variables at the two ends of a delay line. Note that DCFP is called an active device since it contains a clock. DL is called a passive device since it does not contain a clock in the specifications. There is a constraint in writing specifications that is neither passive device nor active device can directly interface to its own kind. If input and output control specifications are added to the front and the end of the above specifications respectively then we get the complete specifications for generating FORTRAN code to simulate the majority logic circuit.

Looking at these specifications it is obvious that they are simple and easy to write. In addition, they are one to one corresponding to the schematic drawing of the circuit diagram. If we have the specifications it should not be difficult to draw out the original schematic circuit diagram. Therefore the specifications not only serve as specifications but also a proper document for the circuit diagram.

The $DCFP(l,CLKn)$ and $DL(i,Xj,Xk)$ are nothing but sub-Hamiltonians of DCFPs and delay lines which can be written down by referring to Fig.6.2 as follows,

$$U_{DCFP} = \frac{\Phi_0 I_m}{2\pi} \cos E \cos x_a \tag{6.1}$$

$$U_{DL} = \frac{(X_a-Y_1)^2}{L_t} + \sum_{\substack{i=1 \\ j \in J}}^{n-1} \frac{(Y_j-Y_{j+1})^2}{2L_t} + \frac{(Y_n-X_b)^2}{2L_t} \tag{6.2}$$

$$K_{DCFP} = \frac{C}{2}(\dot{X}-\dot{X}_e)^2 + \frac{C}{2}(\dot{X}+\dot{X}_e)^2 \tag{6.3}$$

$$K_{DL} = \sum_{\substack{i=1 \\ j \in J}}^{m} \frac{C}{2}\dot{Y}_i^2 \tag{6.4}$$

$$D_{DCFP} = \frac{1}{2R}(\dot{X}-\dot{X}_e)^2 + \frac{1}{2R}(\dot{X}+\dot{X}_e)^2 \tag{6.5}$$

$$D_{DL} = \sum_{\substack{i=1 \\ j \in J}}^{m} \frac{1}{2R_t}\dot{Y}_i^2 \tag{6.6}$$

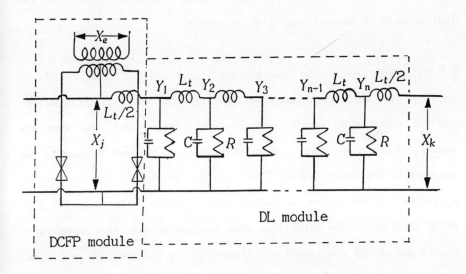

Fig.6.2 A detail circuit diagram showing a DCFP module and a DL module.

$$H_{DCFP} = K_{DCFP} + U_{DCFP} \quad \text{and} \quad H_{DL} = K_{DL} + U_{DL} \tag{6.7}$$

The above expressions can be converted to computer algebra algorithms written in REDUCE(3.0) statements similar to but more elaborated than the algebra algorithms given in the last section. The elaboration are necessary due to the fact that many sub-Hamiltonians are interconnected together instead of a simple isolated system as given in the last example of harmonic oscillation system.

6.5 Implementation of the Complete Specifications

In the last section we mentioned that if input source specifications and the output specifications are included then we get the complete specifications. An input specification is to be used to specify the waveform of the input signal. There are a few input signal sources available in the system. An output specification is for specifying at which location of the circuit we want to observe the flux behaviour. Input and output specifications together with the previous specifications make up the complete specifications for majority logic circuit. Executing these specifications we produce the FORTRAN code. Running the FORTRAN we produce a graphical output. This output is in fact a comprehensive and compact document because the output not only contains the simulation result but also the specifications of the circuit, which can be used to reproduce the circuit configuration when necessary, and circuit parameters with simulation values, which indicate the simulation conditions. Besides the main specifications, there are some auxiliary specifications provided to enhance the system. We shall add in the input, the output and some auxiliary specifications to the previous specifications to illustrate the system as a whole. By gethering them together we obtain the following:

```
% This is the Specification of Majority Logic;
TITLE(MAJOR);
SOURCE(1,XIN0);
SOURCE(2,XIN1);
SOURCE(3,XIN2);
LL(1,XIN0,X1);
LL(2,XIN1,X2);
LL(3,XIN2,X3);
DCFP(1,CK1);
DCFP(2,CK1);
DCFP(3,CK1);
DL(1,X1,X4);
DL(2,X2,X4);
DL(3,X3,X4);
DCFP(4,CK2);
DL(4,X4,0);
OUT3(XIN0,X1,X4);
```

where % is to provide a comment line. TITLE is to give a name to the circuit to be simulated; it would be generated at the output plotting. The first argument of SOURCE(i,XINn) is to specify the DCFP to which the source is being interfaced. The second argument would be used by the system to generate an user input variable in the FORTRAN for designating the initial polarity of the input during simulation. LL is specification for pure inductance coil with similar syntax as DL. Here three LLs are used to connect sources to the buffer DCFPs prior to majority operation.

The three arguments of the OUT3(X_m, X_n, X_p) are for specifying where we want to output the flux for observation(there is another OUT2 for a pair of flux output). Using one OUT3 specification we can output three flux variables for observations. If we want to observe more than three locations we can use more OUT3 specifications then each OUT3 produces output flux for three locations in one piece of output plotting.

There are other device specifications and auxiliary specifications such as LL, RJ2, OPTION and DEFAULT which can be found in the user reference manual<Ohsawa84>.

6.6 Output and Documentation

Based on the above specifications we can obtain output plotting after running the generated FORTRAN program as shown in Fig.6.3.

The upper part of the plotting shows the regenerated input specifications. By looking at these specifications we can easily reconstruct the circuit diagram.

The middle part is the numerical values of the circuit parameters which can either be input at the FORTRAN program running time or input as part of the specifications. If we input a parameter as a specification it would be generated as a default value, that is it would be generated as a constant in the FORTRAN program. If we do not specify it then it would be generated as a FORTRAN runtime input variable.

The lower part of the plotting contains two graphs. The upper graph is a drawing of the three-phase clock for driving the DCFPs. The three phases of the clock are used in the DCFPs specifications as mentioned earlier. The lower graph is the output flux plotting of simulating results. Fig.6.3 was actually generated by OUT3(X2,X3,X4).

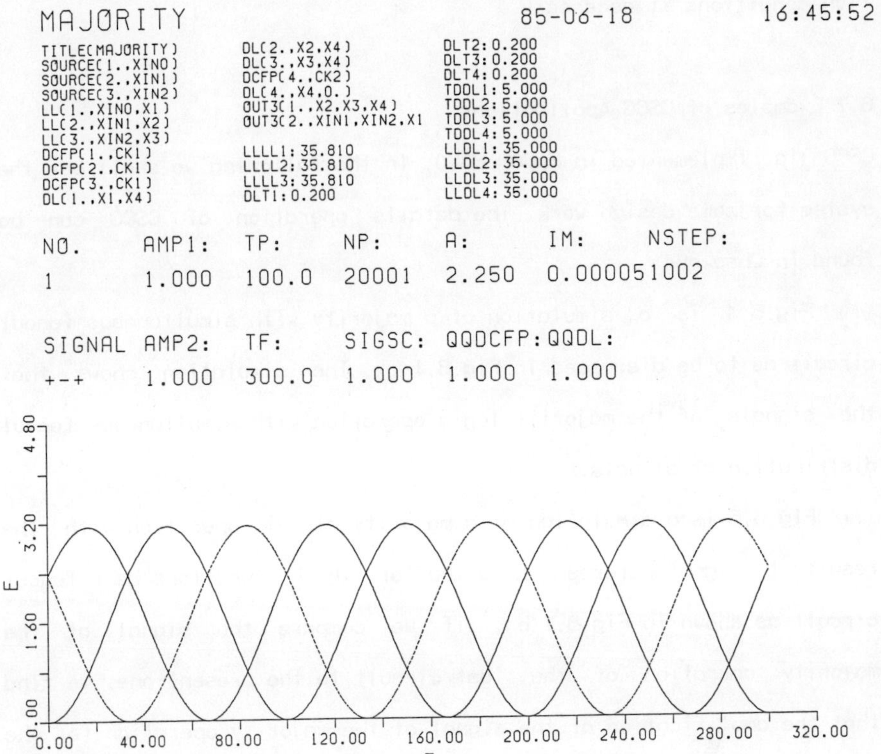
MAJORITY
85-06-18
16:45:52
TITLE(MAJORITY)
SOURCE(1..XIN0)
SOURCE(2..XIN1)
SOURCE(3..XIN2)
LL(1..XIN0,X1)
LL(2..XIN1,X2)
LL(3..XIN2,X3)
DCFP(1..CK1)
DCFP(2..CK1)
DCFP(3..CK1)
DL(1..X1,X4)
DL(2..X2,X4)
DL(3..X3,X4)
DCFP(4..CK2)
DL(4..X4,0.)
OUT3(1..X2,X3,X4)
OUT3(2..XIN1,XIN2,X1
LLLL1: 35.810
LLLL2: 35.810
LLLL3: 35.810
DLT1: 0.200
DLT2: 0.200
DLT3: 0.200
DLT4: 0.200
TDDL1: 5.000
TDDL2: 5.000
TDDL3: 5.000
TDDL4: 5.000
LLDL1: 35.000
LLDL2: 35.000
LLDL3: 35.000
LLDL4: 35.000
NO. AMP1: TP: NP: A: IM: NSTEP:
1 1.000 100.0 20001 2.250 0.000051002
SIGNAL AMP2: TF: SIGSC: QQDCFP: QQDL:
+-+ 1.000 300.0 1.000 1.000 1.000
E
0.00 1.60 3.20 4.80
T
0.00 40.00 80.00 120.00 160.00 200.00 240.00 280.00 320.00
X
-3.00 -1.00 1.00 3.00
CPU TIME = 85.448 SEC
VER H.1.02 85-JUN-14

Fig.6.3 Output plotting which contains the input specifications, input parameter values and the simulation output signals.

The plotting generated by OUT3 contains sufficient information to be kept as document for future reference or reproducing the same simulation conditions if necessary.

6.7 Examples of CSCG Applications

CSCG is implemented in FACOM M380. In the following we shall use the system for some design work. The details operation of CSCG can be found in<Ohsawa84>.

(i) Fig.6.4 is a simulation of a majority with simultaneous fanout circuit as to be discussed in Fig.8.1(a). The simulation shows that the signals of the majority logic operation with simultaneous fanout distribution of signals.

(ii) Fig.6.5 is a simulation of a majority circuit operation with the resultant signal stores in a buffer which functions as a fanout circuit as shown in Fig.8.1(b). If we compare the signal at the majority operation of the last circuit to the present one, we find that the amplification of the signal at the majority operation for the former circuit is somehow smaller due to more circuit loadings.

(iii) Fig.6.6 is a simulation demonstrated the proper operations of a 1/2-Subharmonic Ring Counter which was discussed in the last chapter with Fig.5.14.

(iv) Fig.6.7 is a simulation demonstrated the proper operations of a 1/4-Subharmonic Ring Counter which was discussed in the last chapter with Fig.5.15.

(v) Fig.6.8 is the simulation of one of the two modes (1/6 mode) of the operations of a 1/2-1/6-Subharmonic Ring Counter which was discussed in the last chapter with reference to Fig.5.16. The two modes can be obtained by introducing different combinations of plus and minus signs of δIm noises to the DCFPs, however Fig.5.16 only shows the 1/6-subharmonic mode.

(vi) Fig.6.9 is a simulation on a flip-flop logic circuit which was discussed in chapter 3 with reference to Fig.3.11(a). The simulation shows the proper operation of the flip-flop logic circuit.

(vii) Fig.6.10 is a simulation on a half-adder corresponding to Fig.3.10(b). The specifications given here are slightly different from Fig.3.10(b) but the basic principle is the same. That is in Fig.6.10 additional buffers are added after every majority logic operations as shown in the specifications.

(viii) Fig.6.11 is a simulation on a full-adder circuit which is given in Fig.3.11, also some buffers are added after majority logic operations as shown in the specifications.

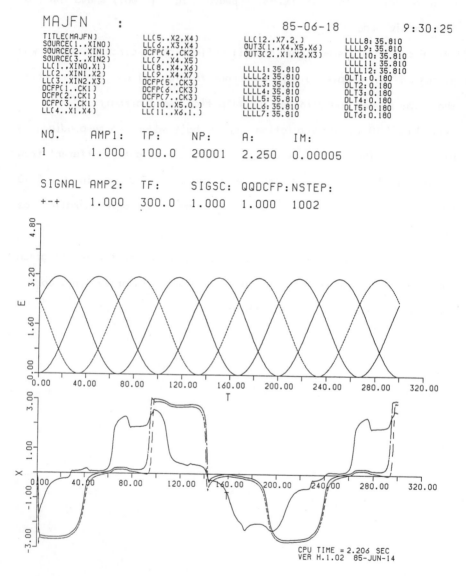

Fig.6.4 Simulation signals of majority logic circuit with three inputs and three fanout outputs without buffer.

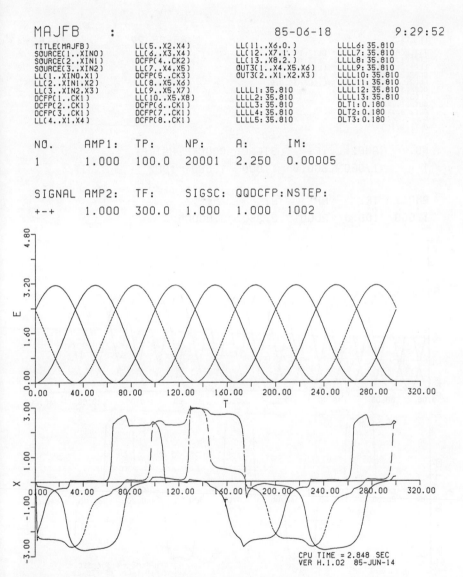

Fig.6.5 Simulation signals of majority logic circuit with three inputs and three fanout outputs connected by a buffer.

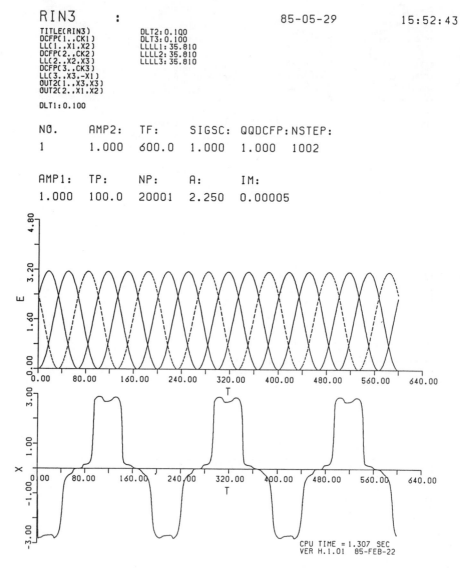

Fig.6.6 Simulation output of a 1/2-subharmonic generator.

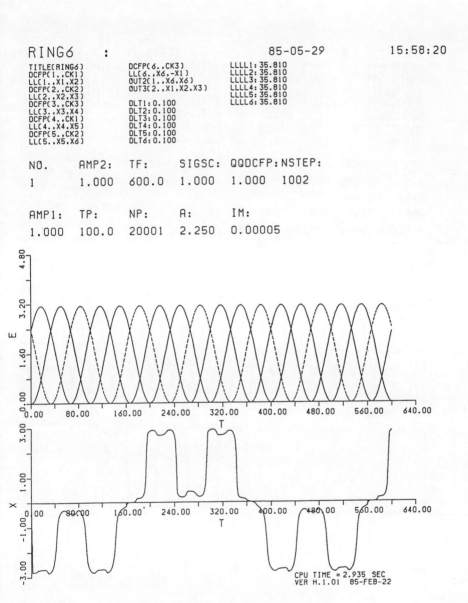

Fig.6.7 Simulation output of a 1/4-subharmonic generator.

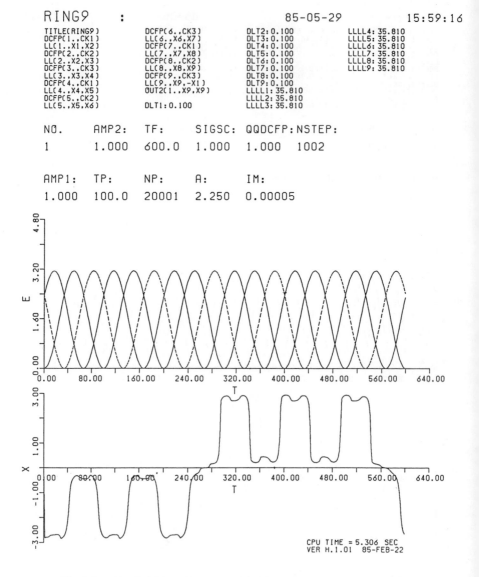

Fig.6.8 Simulation output of 1/6-subharmonic mode for 1/2-1/6-subharmonc generator.

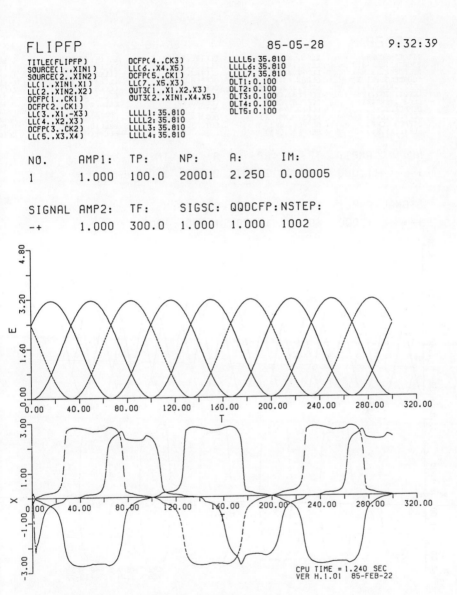

Fig.6.9 Simulation outputs of flip-flop circuit.

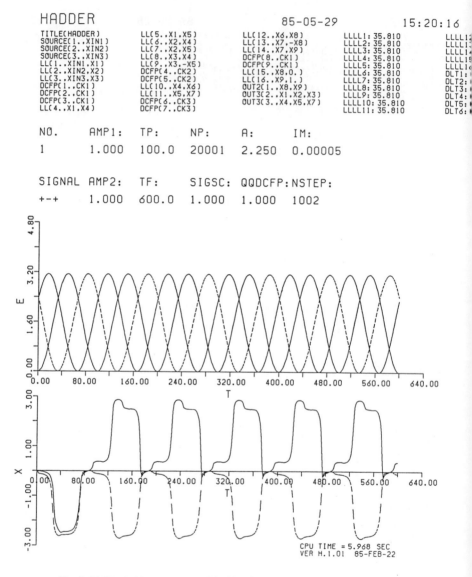

Fig.6.10 Simulation outputs of half-adder.

```
FADDER                                              85-05-29               16:19:34
TITLE(FADDER)       LL(6..X2,X5)      DCFP(8..CK3)      LLLL1: 35.810      LLLL12:
SOURCE(1..XIN1)     LL(7..X2,X6)      DCFP(9..CK3)      LLLL2: 35.810      LLLL13:
SOURCE(2..XIN2)     LL(8..X3,X5)      LL(13..X7,X10)    LLLL3: 35.810      LLLL14:
LL(1..XIN1,X1)      LL(9..X3,X6)      LL(14..X8,X10)    LLLL4: 35.810      LLLL15:
LL(2..XIN2,X2)      DCFP(4..CK2)      LL(15..X9,-X10)   LLLL5: 35.810      LLLL16:
DCFP(1..CK1)        DCFP(5..CK2)      LL(16..X9,X3)     LLLL6: 35.810      LLLL17:
DCFP(2..CK1)        DCFP(6..CK2)      DCFP(10..CK1)     LLLL7: 35.810      DLT1: 0.
DCFP(3..CK1)        LL(10..X4,X7)     LL(17..X10,0.)    LLLL8: 35.810      DLT2: 0.
LL(3..X1,X4)        LL(11..X5,X8)     OUT3(1..X3,X9,X10) LLLL9: 35.810     DLT3: 0.
LL(4..X1,-X5)       LL(12..X6,X9)     OUT3(2..XIN1,X1,X2) LLLL10: 35.810   DLT4: 0.
LL(5..X1,X6)        DCFP(7..CK3)                        LLLL11: 35.810     DLT5: 0.
```

NO.	AMP1:	TP:	NP:	A:	IM:
1	1.000	100.0	20001	2.250	0.00005

SIGNAL	AMP2:	TF:	SIGSC:	QQDCFP:	NSTEP:
++	1.000	600.0	1.000	1.000	1002

Fig.6.11 Simulation outputs of full-adder.

Chapter 7
ANALYSIS OF INDUCTIVE JOSEPHSON LOGIC (IJL)

In this chapter the operational principle of Inductive Josephson Logic(IJL) is analysed starting from a very simple example of a single Josephson junction inductive logic and generalizing to multiple Josephson junctions inductive logics, and thereby, introducing the idea of *forward* and *backward*. Subsequently the idea is used to study various examples of IJL devices.

Inductive Josephson Logic(IJL) was proposed by E.Goto at the RIKEN Symposium<Goto84b>. As compared with the current injection logic(CIL)<Gheewala79> the distinctive differences are:

(i) while CIL uses Josephson junctions as nonlinear resistors to perform combinatorial logic, IJL uses Josephson junctions as nonlinear inductors,

(ii) inputs to CIL are given in the form of dc voltages, inputs to IJL are given in the form of magnetic fluxes,

(iii) output of CIL can be voltage, flux or current, output of IJL can be either flux or current, and

(iv) in relative to CIL the power consumption of IJL is very small owing to the inductive nature of the circuit.

The analysis of IJL in this chapter is based on the Hamiltonian or Lagrangian formalism which was detailed in chapter 4. Also in the course of studying these devices, we shall demonstrate how new devices can be incorporated into the Circuitry Simulation Code Generator(CSCG), which was discussed in the last chapter.

7.1 Potential Energy and Current of IJL

The simplest IJL consisting of a single Josephson junction with an

input flux F_1 and an output flux F_0 are shown in Fig.7.1.
The potential energy of this system is,

$$U = -\frac{J\Phi_0}{2\pi}\cos(\varphi_1-\varphi_0) \qquad (7.1)$$

where $\varphi_k = 2\pi F_k/\Phi_0$, $k=0,1$.
The current going into terminal A_k is denoted by I_k ($k=0,1$) and given by,

$$I_k = \frac{\partial U}{\partial F_k} = \frac{\partial V}{\partial \varphi_k} \qquad (7.2)$$

where $V = 2\pi U/\Phi_0$ and J is the critical current of the Josephson junction.
In the case of the system shown in Fig.7.1 we have,

$$I_1 = \frac{\partial V}{\partial \varphi_1} = -\frac{\partial V}{\partial \varphi_0} = -I_0.$$

This expression as it should be since current has to be conserved.
Now we look into a more complicated system with two inputs φ_1, φ_2, two Josephson junctions with critical currents J_1, J_2 and an output φ_0 as shown in Fig.7.2. Note that the transformer T_0 can be discarded since the winding ratios are 1:1:1. Also all the transformers T_k are ideal flux transformers, that is the mutual inductances are sufficiently large and the leakage inductances are negligible small.
The potential energy of this system is,

$$V = -\sum_{m=1}^{2} J_m \cos\theta_m = -\sum_{m=1}^{2} J_m \cos\left(\sum_{k=0}^{2} w_{mk}\varphi_k\right) \qquad (7.3)$$

where

$$\theta_1 = \varphi_0 + \frac{\varphi_1}{2} + \varphi_2 = \sum_{k=0}^{2} w_{mk}\varphi_k \Big|_{m=1}$$

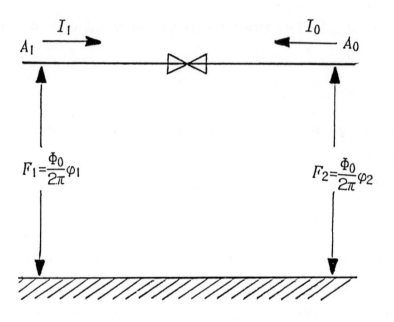

Fig.7.1 A simplest IJL circuit consists of a Josephson junction with an input flux F_1 and an output flux F_0.

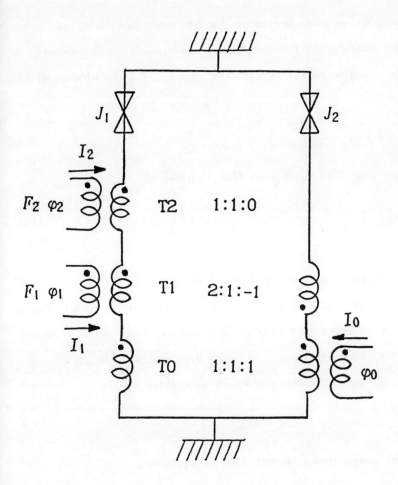

Fig.7.2 A two inputs and one output IJL circuit with two Josephson junctions.

$$\theta_2 = \varphi_0 - \frac{\varphi_1}{2} = \sum_{k=0}^{2} w_{mk}\varphi_k \big|_{m=2}$$

are the respective phase angles of the Josehpson junctions. (w_{mk}) is a matrix representing the ratios of the transformer winding numbers to be called winding matrix hereinafter. For Fig.7.2 the winding matrix is,

$$W = (w_{mk}) = \begin{bmatrix} 1 & 1/2 & 1 \\ 1 & -1/2 & 0 \end{bmatrix} \tag{7.4}$$

The current associated with each flux is given by,

$$I_k = \frac{\partial V}{\partial \varphi_k} \qquad k=0,1,2 \tag{7.5}$$

7.2 Formulations of General Form of IJL Circuit

General form of IJL to be treated consists of K input fluxes $\varphi_{k'}$ ($k'=1\ldots K$), an output flux φ_0, M Josephson junctions with respective critical current J_m ($m=1\ldots M$) and a winding matrix $W=(w_{mk})$. The potential energy of the system can be written as,

$$V = -\sum_{m=1}^{M} J_m \cos\theta_m = -\sum_{m=1}^{M} J_m \cos\left(\sum_{k=0}^{K} w_{mk}\varphi_k\right) \tag{7.6}$$

The current associated with each flux is given by

$$I_k = \frac{\partial V}{\partial \varphi_k} \qquad k=0\ldots K. \tag{7.7}$$

The input fluxes can be supplied either by dc SQUID circuits or DCFP and the output can be connected to dc SQUID or DCFP as well. In this book we shall treat the case in which the output is a DCFP and some of the inputs are DCFPs and some may be connected to dc flux biases. Without loss of generality we assume φ_K represents dc bias whenever dc bias is to be used.

Considering the case of DCFP connected to the IJL is excited then output logic flux level of DCFP could be of any arbitrary fixed value F_L which again can be transformed into the desirable values for IJL input by flux transformer. For the sake of simplicity and again without loss of generality we assume that $F_L=\pm\Phi_0/2$ so that $\varphi_k=\pm\pi$. Also assume that $\varphi_k=\pi$ represents 'true' logic value and $\varphi_k=-\pi$ represents 'false' logic value respectively. Now we are ready to introduce the idea of *forward* and *backward* for the subsequent IJL circuit analysis.

Forward is a state in which the DCFPs connected to the inputs of IJL are excited so that $\varphi_{k'}\sim\pm\pi$, and the DCFP connected to the output is not excited so that $|\varphi_0|\ll\pi$.

We denote the I_k of the *forward* by

$$IF_k(\varphi_{k'}) = I_k(0,\varphi_{k'}). \qquad (7.8)$$

Note that $IF_0(\varphi_{k'})$ is the *forward* output current which is used as the seed current to the DCFP connected to the output side of IJL. Therefore $IF_0(\varphi_{k'})$ should have definite signs which depend on $\varphi_{k'}$ and represent a specific logic function. In addition it is desirable to have $IF_0(\varphi_{k'})$ taking the following values for all the possible input combinations of $\varphi_{k'}$, that is,

$$IF_0 = \pm J_L, \qquad (7.9)$$

where J_L is the logic seed current at the output of IJL. We call (7.9) the homogeneous output current conditions.

$IF_{k'}$ is the *forward* input current associated with the k'th input flux $\varphi_{k'}$. It is also desirable to have the absolute values of $IF_{k'}$ to be small for all possible combinations of inputs, since $IF_{k'}$ is the *forward* loading current of the DCFP at the input side of the IJL.

Next we shall consider *backward*. *Backward* is a state in which the DCFP connected to the inputs are not excited so that $|\varphi_{k'}| \ll \pi$ (in the case of φ_K is dc bias, $\varphi_K = \pi$ is an exception), and the DCFP connected to the output is excited so that $\varphi_0 \sim \pm \pi$.

Now we define *backward* current as,

$$IB_k(\varphi_0) = I_k(\varphi_0, all\ \varphi_{k'} = 0\ except\ dc\ bias\ \varphi_K = \pi) \qquad (7.10)$$

Note that IB_0 is the *backward* current at the DCFP connected to output and $IB_{k'}$ s are the *backward* currents at the DCFP connected to inputs of IJL. Hence it is desirable to keep the absolute values of all the *backward* currents IB_k small thus reducing the loading current to the DCFP at the output side as well as reducing the undesirable reaction current to the DCFP at the input side.

In addition to the *backward* current we can define a *backward* transient current which is a current comes from the transition from *forward* to *backward* given by,

$$IB*_k(\varphi_0) = I_k(0 < |\varphi_0| < \pi, all\ \varphi_{k'} = 0\ except\ dc\ bias\ \varphi_K = \pi) \qquad (7.11)$$

In most of the cases the absolute maximum transient current occurs at $\varphi_0 = \pi/2$.

In the next section we shall give examples to show how all these requirements can be realized in some of the IJL circuits.

Another important characteristics of IJL useful for comparing the quality of various IJL circuits is the tolerance function. The tolerance function f is defined as the variation of output *forward* current with respect to the simultaneous variations of junction critical current δJ_m and phase angles $\delta \varphi_{k'}$ as given by,

$$\delta IF_0 = f(\delta J_m, \delta \varphi_{k'}) \qquad (7.12)$$

7.3 Examples of IJL Circuits

Following are some examples illustrating what we have discussed earlier. In these examples, the values of the arguments given to IF_k and IB_k are $\pm\pi$ and given to $IB*_k$ are $\pm\pi/2$ respectively.

Example 1

This is a circuit called RJ2(Repeater with Junction 2 in number). Fig.7.2 is a circuit realizing RJ2 in which φ_2 is dc bias thus $\varphi_2=\pi$, and $J_1=J_2=J/2$.

According to (7.6) the potential energy of RJ2 is given by,

$$V = -J(\cos(\varphi_0 + \frac{\varphi_1}{2}+\pi) + \cos(\varphi_0 - \frac{\varphi_1}{2})) = -2J\sin\varphi_0\sin\frac{\varphi_1}{2} \quad (7.13)$$

Thus the currents are

$$I_0 = \frac{\partial V}{\partial \varphi_0} = -2J\cos\varphi_0\sin\frac{\varphi_1}{2}, \quad (7.14)$$

$$I_1 = \frac{\partial V}{\partial \varphi_1} = -J\sin\varphi_0\cos\frac{\varphi_1}{2}. \quad (7.15)$$

From the above we get $IF_0(\pm\pi) = \mp 2J$ which satisfy the conditions of homogeneous current at output side given by (7.9). Also we get $IF_1(\pm\pi)=0$, $IB*_0(\pm\pi/2)=0$, $IB*_1(\pm\pi/2)=\pm J$, $IB_0(\pm\pi/2)=0$ and $IB_1(\pm\pi)=0$, thus all the undesirable current are ideally minimized. The tolerance function is,

$$\delta IF_0(\varphi_k\cdot) = 2\delta J + \delta\varphi_1 \quad (7.16)$$

We shall compare this with the next example to see the relative tolerances of this IJL circuit.

Example 2

RJ1 is a circuit after removing a Junction and the input bias from

RJ2. The potential energy and the currents of RJ1 are as follows,

$$V = -J\cos(\varphi_0 + \frac{\varphi_1}{2}) \quad (7.17)$$

$$I_0 = J\sin(\varphi_0 + \frac{\varphi_1}{2}) \quad (7.18)$$

$$I_1 = \frac{I_0}{2} \quad (7.19)$$

From the above we get $IF_0(\pm\pi) = \pm J$, the homogeneous output conditions are satisfied. But $IF_1 = \pm J/2$, $IB*_0 = \pm J/2$, $IB*_1 = \pm J/2$, $IB_0 = 0$, and $IB_1 = 0$, make RJ1 inferior to RJ2 for having nonzero IF_1.

The tolerance function of RJ1 is,

$$\delta IF_0 = \delta J + \frac{J}{2}\delta\varphi_1 \quad (7.20)$$

which is the same as RJ2.

Example 3

This is a circuit called H4J4<Goto84b>. The potential energy of this circuit is,

$$V = -J\Big(\cos(\varphi_0 + \frac{\varphi_1}{2} + \frac{\varphi_2}{2} + \frac{\varphi_3}{2}) + \cos(\varphi_0 - \frac{\varphi_1}{2} + \frac{\varphi_2}{2} - \frac{\varphi_3}{2}) + \cos(\varphi_0$$
$$+ \frac{\varphi_1}{2} - \frac{\varphi_2}{2} - \frac{\varphi_3}{2}) + \cos(\varphi_0 - \frac{\varphi_1}{2} - \frac{\varphi_2}{2} + \frac{\varphi_3}{2})\Big) \quad (7.21)$$

$$= 4J\Big(\cos\varphi_0\cos\frac{\varphi_1}{2}\cos\frac{\varphi_2}{2}\cos\frac{\varphi_3}{2} + \sin\varphi_0\sin\frac{\varphi_1}{2}\sin\frac{\varphi_2}{2}\sin\frac{\varphi_3}{2}\Big).$$

Thus we get

$$IF_0 = J\sin\frac{\varphi_1}{2}\sin\frac{\varphi_2}{2}\sin\frac{\varphi_3}{2} = \pm 4J, \quad (7.22)$$

which is a parity function and,

$IF_{k'}=0$, $IB*_k=0$, $IB_k=0$,

$$\delta IF_0 = 4\delta J - 2J\left((\delta\varphi_1)^2 + (\delta\varphi_2)^2 + (\delta\varphi_3)^2\right) \tag{7.23}$$

Example 4

This is a circuit called P3J1 (Parity of 3 bits with Junction 1 in number). The potential energy is given by,

$$V = -J\cos\left(\varphi_0 + \frac{\varphi_1+\varphi_2+\varphi_3}{2}\right) \tag{7.24}$$

thus,

$$IF_0 = -J\sin\left(\frac{\varphi_1+\varphi_2+\varphi_3}{2}\right) = \pm J, \quad IF_{k'}=\pm J/2, \quad IB*_0=\pm J, \quad IB_{k'}=\pm J/2, \quad IB_k=0$$

This circuit is inferior to H4J4 since $IF_{k'}$ is not zero. Also if we assume $\delta\varphi_{k'}=\delta\varphi$ for all k', then tolerance function of P3J1 is larger than H4J4, since tolerance function of P3J1 is given by,

$$\delta IF_0 = \delta J \pm \frac{J}{2}(\delta\varphi_1+\delta\varphi_2+\delta\varphi_3)^2 = \delta J \pm \frac{9J}{2}(\delta\varphi)^2 \tag{7.25}$$

whereas tolerance function of H4J4 is given by,

$$\delta IF_0 = \delta J \pm \frac{J}{2} = \delta J \pm \frac{3J}{2}(\delta\varphi)^2 \tag{7.26}$$

Example 5

S2J4 is a circuit with its name derived from Selector of 2 bits with Junctions 4 in number. The potential energy of this circuit is,

$$V = -J\Big(\cos(\varphi_0+\frac{\varphi_1}{2}+\frac{\varphi_3}{4}+\frac{\varphi_4}{4}) + \cos(\varphi_0-\frac{\varphi_1}{2}+\frac{\varphi_3}{4}+\frac{\varphi_4}{4}+\pi) + \cos(\varphi_0+\frac{\varphi_2}{2}$$
$$+\frac{\varphi_3}{4}-\frac{\varphi_4}{4}) + \cos(\varphi_0-\frac{\varphi_2}{2}+\frac{\varphi_3}{4}-\frac{\varphi_4}{4}+\pi)\Big) \tag{7.27}$$

Thus we get

$IF_0=\pm 2J$, $IB*_{1,2}=\pm 2J$, $IB*_{0,3,4}=0$, $IF_{1,2}=0$ $Max(|IF_{3,4}|)=J/2$, $IB_k=0$.

It can be shown by a truth table that this circuit realizes the following logic,

$$x_0 = (x_1 \& (x_3 \oplus x_4)) \cup (x_2 \& (\overline{x_3 \oplus x_4})) \tag{7.28}$$

where x_k is the logical variable represented by φ_k.
The tolerance function is,

$$\delta IF_0 = \delta J + \frac{J}{4}(\delta\varphi_3 + \delta\varphi_4) \pm \frac{J}{2}((\delta\varphi_1)^2 + (\delta\varphi_2)^2) \tag{7.29}$$

7.4 Incorporating IJL Modules Into CSCG for Simulations

Forward and *backward* which we have discussed are the two steady states of the IJL operation. For the dynamic transition between *forward* and *backward*, simulation should be done to determine the proper values of the circuit parameters such as the inductance values for IJL and DCFP coupling, and the ratio of junction critical current of IJL to DCFP to be called α.

We have pointed out in chapter 6 that CSCG is an extensible circuit simulation system and new logic functions can be added into it easily. We shall add some of the useful logic components being discussed earlier into the generator, and in a way, also demonstrate the extensibility of the CSCG.

The general potential energy of IJL is given by (7.6) multiplying by $\Phi_0/2\pi$, the kinetic energy and the dissipating function are respectively given by

$$K = \sum_{m=1}^{M} \frac{C_m}{2} (\sum_{k=0}^{K} w_{mk} \dot{F}_k)^2, \tag{7.30}$$

$$D = -\sum_{m=1}^{M} \frac{1}{2R_m} (\sum_{k=0}^{K} w_{mk} \dot{F}_k)^2, \tag{7.31}$$

where C_m and R_m are the capacitors and resistors of the equivalent Josephson junctions.

Given the potential energy, kinetic energy and the dissipating function we shall be able to derive the Lagrangian and Hamiltonian which are necessary for the design of algorithms for new logic components to be added into the generator. We shall define the syntax for some of the useful examples given above and make them as part of the semantic in CSCG for writing specifications.

To simplify the design of IJL module, we assume the DCFPs are connected via pure inductance lines to the inputs of IJL but the output of IJL is directly connected to a DCFP. With such a partition the DCFP at the output side of IJL is included as part of the IJL module, but the DCFPs connected to the input side of IJL are excluded from this module.

Following are the respective REDUCE statement for specifying RJ2 and S2J4,

$\quad$ RJ2(n, CKi), $\qquad$ S2J4(n, CKi),

where n is an integer to specify a particular IJL in a circuit configuration, CKi is the clock used to drive the DCFP connected at the output side of IJL. Since the DCFP at output side is included as part of the IJL module, therefore the potential energy of the RJ2 and S2J4 modules are given by adding the potential energy of the DCFP connecting to their respective output side to their own potential energy. RJ2(n, CKi) statement would generate An as an interface variable at input side, and Xn as an interface variable at output side of RJ2 module. The S2J4(n, CKi) would generate An, Bn, Sn, Tn as interface variables at input side and Xn as an interface variable at output side of S2J4. Therefore the interface variables and the kinetic energy of these two examples are as follows,

(i) For RJ2 let $X_n = \varphi_0$, $A_n = \varphi_1^+$, where φ_0 is the interface flux of IJL output and DCFP, and φ_1^+ is the interface flux of inductance coil line and IJL input.

The kinetic energy of RJ2 is,

$$K = \frac{C\Phi_0^2}{8\pi^2}(\dot{\varphi}_0^2 + \alpha(\dot{\theta}_1^2 + \dot{\theta}_2^2))$$

$$= \frac{C\Phi_0^2}{8\pi^2}(\dot{\varphi}_0^2 + \alpha((\dot{\varphi}_0 - \frac{\dot{\varphi}_1}{2})^2 + (\dot{\varphi}_0 + \frac{\dot{\varphi}_1}{2})^2))$$

$$= \frac{C\Phi_0^2}{8\pi^2}((1+2\alpha)\dot{\varphi}_0^2 + \frac{\alpha}{2}\dot{\varphi}_1^2) \qquad (7.32)$$

(ii) For S2J4 let $X_n = \varphi_0$, $A_n = \varphi_1^+$, $B_n = \varphi_2^+$, $S_n = \varphi_3^+$, and $T_n = \varphi_4^+$, where φ_0 is the interface flux of IJL output and DCFP, and φ_k^+ is the interface flux of inductance coil line and IJL inputs. Thus the kinetic energy is,

$$K = \frac{C\Phi_0^2}{8\pi^2}(\dot{\varphi}_0^2 + \alpha(\dot{\theta}_1^2 + \dot{\theta}_2^2 + \dot{\theta}_3^2 + \dot{\theta}_4^2))$$

$$= \frac{C\Phi_0^2}{8\pi^2}(\dot{X}_0^2 + \alpha(4(\dot{X}_0 + \frac{\dot{X}_3}{4})^2 + \frac{\dot{X}_1^2}{2} + \frac{\dot{X}_2^2}{2} + \frac{\dot{X}_4^2}{4})) \qquad (7.33)$$

(i) Fig.7.3 is a simulation result of RJ2 circuit using the CSCG system. The waveforms of the flux are in the good shape. However experiments shall be done to confirm the simulation.

(ii) Fig.7.4 is a simulation of S2J4 circuit using the CSCG system. The waveforms of the flux are also in good shapes and experiments shall also be done to confirm the simulation.

```
RJ2S2         :                              85-06-17         16:52:08
TITLE(RJ2S2)         OUT2(2..X1.X3)
SOURCE(1..XIN1)
LL(1..XIN1.X1)       LLLL1: 35.810
DCFP(1..CK1)         LLLL2: 35.810
LL(2..X1.A2)         LLLL3: 35.810
RJ2(2..CK2)          LLLL4: 35.810
LL(3..X2.0.)         LLLL5: 35.810
LL(4..X2.X3)         DLT1: 0.000
DCFP(3..CK3)         DLT2: 0.000
LL(5..X3.0.)         DLT3: 0.000
OUT3(1..XIN1.A2.X2)

NO.   AMP1:   TP:     NP:     A:      IM:         NSTEP:
1     1.000   100.0   20001   2.250   0.000051002

SIGNAL  AMP2:   TF:     SIGSC:  QQDCFP: ALPHA:
+       1.000   300.0   1.000   1.000   0.600
```

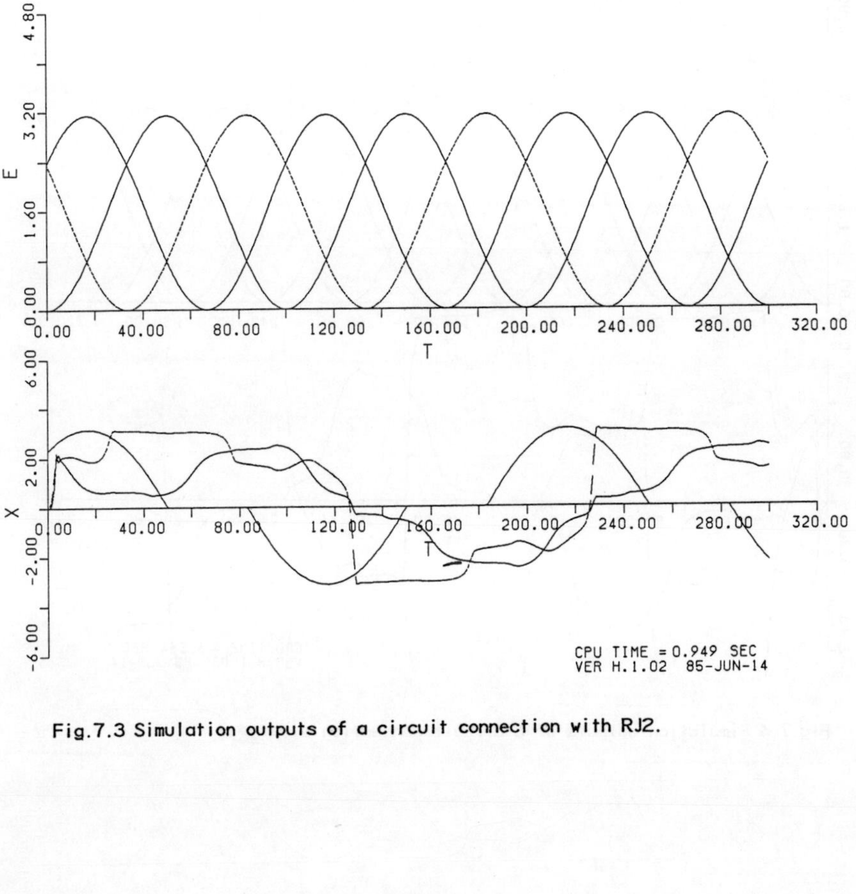

CPU TIME = 0.949 SEC
VER H.1.02 85-JUN-14

Fig.7.3 Simulation outputs of a circuit connection with RJ2.

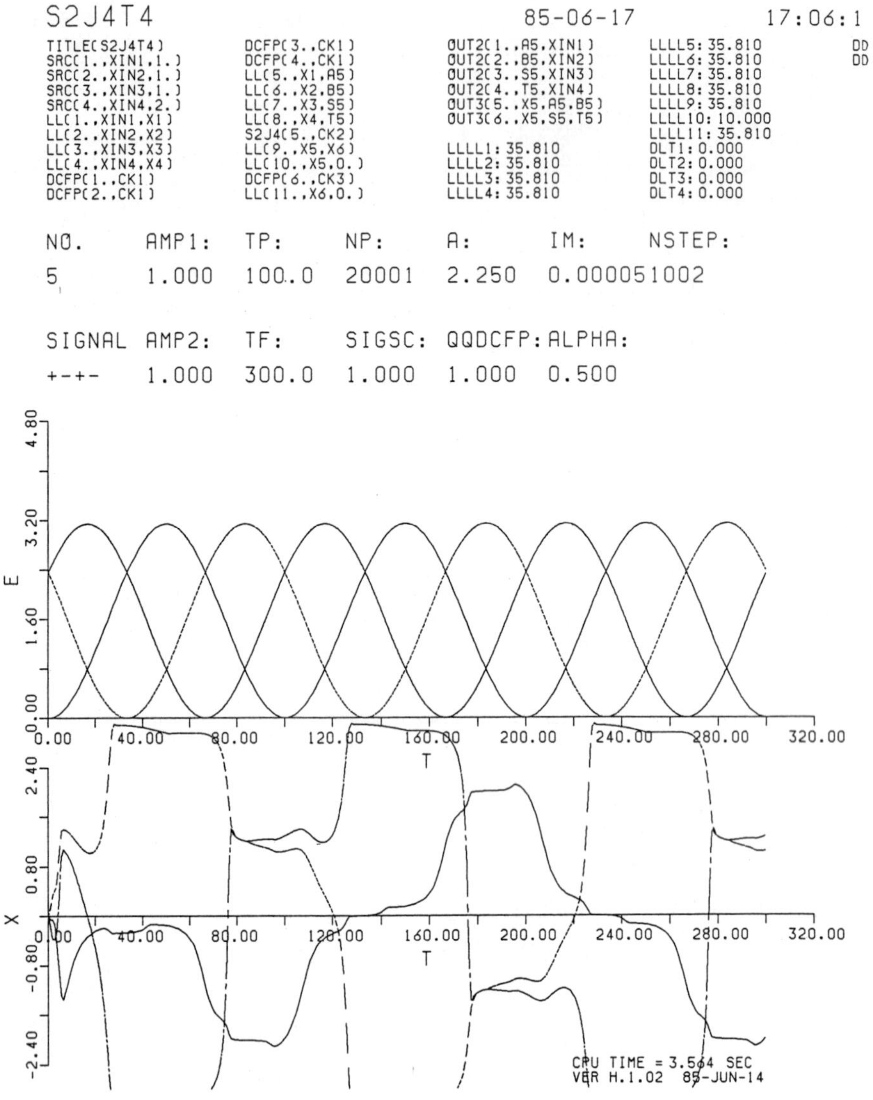

S2J4T4
85-06-17
17:06:1
TITLE(S2J4T4)
SRC(1.,XIN1,1.)
SRC(2.,XIN2,1.)
SRC(3.,XIN3,1.)
SRC(4.,XIN4,2.)
LL(1.,XIN1,X1)
LL(2.,XIN2,X2)
LL(3.,XIN3,X3)
LL(4.,XIN4,X4)
DCFP(1.,CK1)
DCFP(2.,CK1)
DCFP(3.,CK1)
DCFP(4.,CK1)
LL(5.,X1,A5)
LL(6.,X2,B5)
LL(7.,X3,S5)
LL(8.,X4,T5)
S2J4(5.,CK2)
LL(9.,X5,X6)
LL(10.,X5,0.)
DCFP(6.,CK3)
LL(11.,X6,0.)
OUT2(1.,A5,XIN1)
OUT2(2.,B5,XIN2)
OUT2(3.,S5,XIN3)
OUT2(4.,T5,XIN4)
OUT3(5.,X5,A5,B5)
OUT3(6.,X5,S5,T5)
LLLL1: 35.810
LLLL2: 35.810
LLLL3: 35.810
LLLL4: 35.810
LLLL5: 35.810
LLLL6: 35.810
LLLL7: 35.810
LLLL8: 35.810
LLLL9: 35.810
LLLL10: 10.000
LLLL11: 35.810
DLT1: 0.000
DLT2: 0.000
DLT3: 0.000
DLT4: 0.000
DD
DD
NO. AMP1: TP: NP: A: IM: NSTEP:
5 1.000 100.0 20001 2.250 0.000051002
SIGNAL AMP2: TF: SIGSC: QQDCFP: ALPHA:
+-+- 1.000 300.0 1.000 1.000 0.500
E
4.80
3.20
1.60
0.00
0.00 40.00 80.00 120.00 160.00 200.00 240.00 280.00 320.00
T
X
2.40
0.80
-0.80
-2.40
0.00 40.00 80.00 120.00 160.00 200.00 240.00 280.00 320.00
T
CPU TIME = 3.564 SEC
VER H.1.02 85-JUN-14

Fig.7.4 Simulation outputs of a circuit connection with S2J4.

Chapter 8
COMPUTER TECHNOLOGY BASED ON DCFPs

Based on the discussion in chapter 3 about the majority logic circuit it is known that we can design a digital computer system using majority logic with an additional NOT logic. In chapter 6 simulation was done to confirm that the DCFP can be used to make a majority logic. The operation of DCFP is experimentally verified together with NOT logic realization using a flux inverting transformer<Harada85a>. With these basic logic circuits then, in principle, it is possible to design a DCFP digital computer system. However in actual designing a digital system, additional problems such as fanin, fanout and coupling problems have to be considered. These problems are common to any digital system design but the solution to them are different for different technology concerned. We shall study these problems in DCFP digital system design and find out how these problems in DCFP digital system should be handled.

Logic circuit alone does not make a complete computer system, we need memory to store information or data to be processed. The main memory built from the DCFP was also invented by E.Goto<E.Goto84c>. To have an overall understanding on how DCFP can be used to design the high speed computer we shall also briefly discuss the DCFP memory unit and processor. Cyclic Pipeline Machine(CPM)<Shimizu84, Goto84d> was proposed to take full advantages of the DCFP characteristics. Shimizu <Shimizu85> has given detailed account on CPM. As a last chapter to conclude up the DCFP computer technology, we shall also touch on DCFP memory and CPM later in this chapter.

8.1 Constraints and Wiring Rules

The basic logic device being used to build the parametron computer was given in chapter 3. As DCFP operates on the similar principle as conventional parametron, it is natural to expect that all the basic digital logics for computer design such as AND, OR, NOT, FLIP-FLOP, ADDER, COUNTER, REGISTER shall be constructed similar to the conventional parametron technology. This is generally true, yet there are some constraints that make the logic circuits as well as the overall architecture of a DCFP computer somehow different from the conventional parametron computer. One of the difference is due to the smaller amplification obtainable from the DCFP which limits the number of fanin and fanout achievable by the DCFP.

From our simulation in section 6.7 we found that if fanout and majority operations are performed by the same DCFP as shown in Fig.8.1(a), then the level of the amplified signal becomes much smaller, and it is vulnerable to various noises. Therefore a more reliable circuit is to have the majority logic operations and the fanout operations to be done separately at different clock cycles as shown in Fig.8.1(b). In Fig.8.1(b) the left-hand-side is a majority logic circuit and the circuit at right-hand-side which mirrors the majority logic configuration, serves only as a fanout circuit to distribute signal to the next stage. The consequence of this is more clock cycles are needed to operate a logic circuit as compared with the conventional parametron logic circuit. Since DCFP is operate on very high speed then the reliability we gain by this configuration off set the penalty.

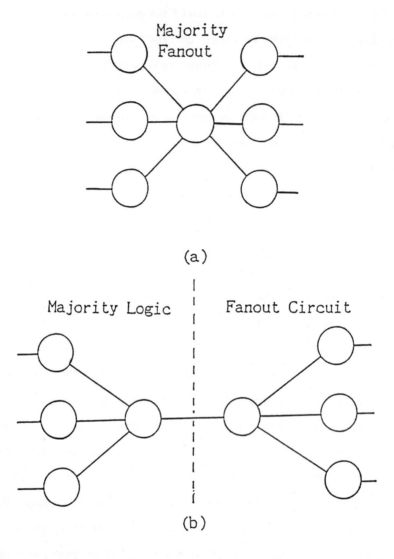

Fig.8.1 (a) connection of majority logic circuit and fanout circuit without buffer, (b) connection of majority circuit and fanout circuit via a DCFP buffer.

Switching speed and propagation delay also are the constraints which affect the wiring rule and packaging. Because of the high switching speed and the propagation delay caused by the dielectric and London's penetration the wiring between any two DCFPs must be kept within $1/6cm$. Thus conventional hierarchical packaging of chips, cards and back-panels cannot meet the "wiring rules". The three dimensional packaging called *Antenna Method* was proposed by E.Goto<Goto84d>. Using this method a CPU composed of a million DCFPs will be packed in 1 cm^3 without heat sink problems since one DCFP consumes only 1 nW of power.

8.2 Impossibility of Directional Coupling

When coupling DCFPs together to make more complicated logic circuit we shall aware of the fact that DCFP is a two ports device with bilateral signals. Such a device would not be able to have a directional neat flow of current. This fact is known as a reciprocity theorem which states that the forward and the backward flow of currents between two DCFPs coupling by linear inductance(or transformer) are nearly equal. To explain this we refer to Fig.8.2 where the box denotes some linear inductance circuit(or transformer), and A and B are two DCFPs coupled by the inductances. The forward current I_1 and flux Φ_1 are related to the backward current I_2 and flux Φ_2 as follows,

$$\begin{bmatrix} \Phi_1 \\ \Phi_2 \end{bmatrix} = \begin{bmatrix} L_1 & M \\ M & L_2 \end{bmatrix} \begin{bmatrix} I_1 \\ I_2 \end{bmatrix} \qquad (8.1)$$

Inversely the current flows can be written in terms of fluxes as follows,

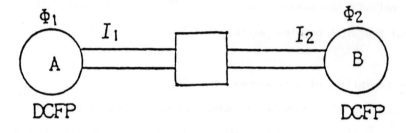

Fig.8.2 DCFPs coupling with linear inductances which are denoted by a box.

$$\begin{bmatrix} I_1 \\ I_2 \end{bmatrix} = \begin{bmatrix} \Lambda_1 & \Lambda_{12} \\ \Lambda_{21} & \Lambda_2 \end{bmatrix} \begin{bmatrix} \Phi_1 \\ \Phi_2 \end{bmatrix} \tag{8.2}$$

where

$$\begin{bmatrix} \Lambda_1 & \Lambda_{12} \\ \Lambda_{21} & \Lambda_2 \end{bmatrix} = \begin{bmatrix} L_1 & M \\ M & L_2 \end{bmatrix}^{-1} \tag{8.3}$$

If Φ_1 is excited but Φ_2 is not then we have $\Phi_1 = \Phi_{out}$ and $\Phi_2 \sim 0$. Owing to the symmetry of the matrix we get, $I_2 = \Lambda_{21}\Phi_{out} = \Lambda_{12}\Phi_{out} = I_1$. Thus directional flow of current is impossible since the forward and backward flows of currents are equal.

8.3 Relay Noise

Owing to the backward flow of current there is a so called relay noise problem which has to be taken into consideration when making circuits out from DCFPs. Relay noise is a more general phenomenon not only exists in the DCFP connections but also exists in other connection of bilateral circuit elements as long as these circuits are connected in three stages driven by three phases of clock. Sometimes we also call the relay noise in parametron as grandchildren noise since the noise is caused by the backward flow of current from the third stage to the first stage of a circuit relayed by the device in the second stage. Though we cannot eliminate the backward flow current by linear inductance coupling owing to the reciprocity current relation, however we can reduce or eliminate the relay noise by the three schemes to be discussed later. Now we shall make more details study on the relay noise.

Fig.8.3 shows a connection of three stages DCFP with m branches of fanout from each DCFP thus there are totally m^2 DCFPs in the

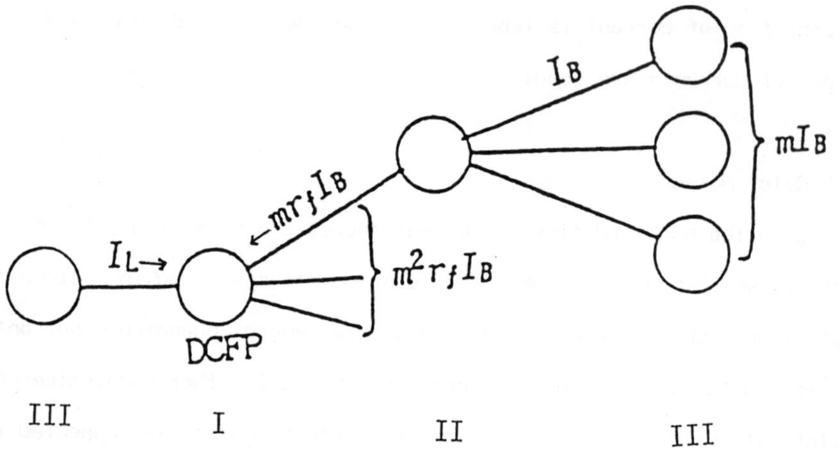

Fig.8.3 Relay noise flowing backward from a DCFP fanout circuit connection.

third(phase III)stage. Let I_B be the backward flow current and r_f be the relay factor which is defined in Fig.8.4 as $r_f=I_{B21}/I_{B32}$.
Then the total relay noise is related to the forward loading current I_L by $I_L=m^2 r_f S_r I_B$, where S_r is a safety factor which could ensure a net forward flow of current if its value is larger than unity. According to the reciprocity theorem as discussed earlier we get $I_L=I_B$ and thus $S_r=1/(m^2 r_f)$. For a simple circuit of inductance coupling as shown in Fig.8.5 the relay factor is given by,

$$r_{f0}=(\frac{2(1\pm\delta)}{L_J}+\frac{2}{L_c})^{-1}\frac{1}{L_c} \doteqdot \frac{L_J(1\pm\delta)}{2L_c} \qquad \text{for} \quad L_c \gg L_J \qquad (8.4)$$

where δ is the percentage of δI_m noise. The relay factor of this circuit can be improved by Fig.8.6 circuit configuration to be called the cancellation coupling scheme, which is constructed by adding a shunt inductance L_r to feed with reversing current produced by a transformer coupling to counteract the relay noise. In this case the relay factor becomes $r_f=\delta r_{f0}$.
The drawback of this scheme is that additional m^2 wirings are required for most of the circuit applications, however a special application in which the circuit is used as a two stages booster to drive logic circuit then only one of this connection is required as shown in Fig.8.7 by the dotted line.
Alternatively we can use a hybrid transformer coupling scheme to eliminate the undesirable relay noise. A hybrid transformer is used in telephone set to eliminate the speaker's voice to be fedback to the earpiece from the mouthpiece. In DCFP it is realized as in Fig.8.8 for eliminating the relay noise.
Relay noise cannot be totally eliminated due to the uncertainty of the

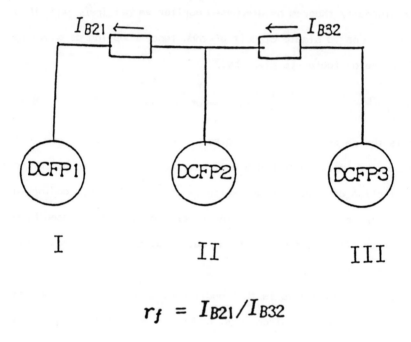

Fig.8.4 Schematic illustration of the definition of relay noise.

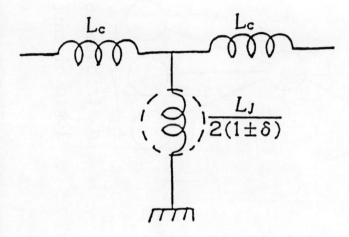

Fig.8.5 Schematic illustration of a simple relay factor.

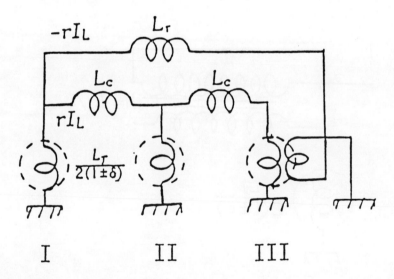

Fig.8.6 Circuit for cancellation coupling scheme to reduce the relay noise.

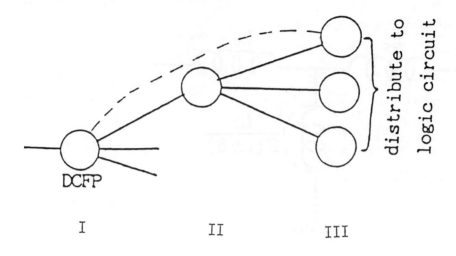

Fig.8.7 Booster circuit with a dotted line to realize a cancellation coupling scheme for DCFP signal fanout distribution.

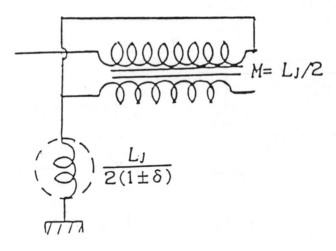

Fig.8.8 Circuit for hybrid trnasformer coupling scheme to reduce the relay noise.

deviation of the critical current δIm. Thus the relay factor is reduced in proportion to this uncertainty, that is $r=\delta r_{fo}$.

In comparing the resource used in the two schemes the former scheme requires m^2 inductance coupling for most of the circuit application but it requires only one inductance coupling if the fanout circuit is to be used as a two stages booster for some logic circuit. Whereas in the latter scheme, regardless of the application m hybrid transformers are always required.

The third method to reduce relay noise is to used one of the IJL logic circuits known as RJ2, which was studied in chapter 7, for logic circuit fanout connection. Assuming that a signal αI_m input to a DCFP is fanout to m branch and the fanout can consecutively be done for three stages as shown in Fig.8.9, where a block denotes a RJ2 module which includes a DCFP connected to the output. The number of fanout m to be expressed in terms of three parameters namely a safety factor s, a current ratio α and a phase angle, can be derived by considering from the last stage of the fanout in Fig.8.9 and working backward. The *forward* current at phase III clock can be written as follows,

$$mIF_1 = m\alpha I_m \sin\varphi_{III} \cos\frac{\varphi_{II}}{2} \tag{8.5}$$

$$\frac{2\pi}{\Phi_0}\Phi_{II} = mIF_1 \cdot \frac{L_J}{2} \quad i.e. \quad \varphi_{II} = \frac{m\alpha}{2}\sin\varphi_{III}, \tag{8.6}$$

where $\cos\varphi_{II}/2 \sim 1$. Similarly the *forward* current in phase II clock stage is,

$$mIF_1 = \frac{2\alpha m I_m}{2}\sin\varphi_{II}\cos\frac{\varphi_I}{2} \tag{8.7}$$

Since $\sin\varphi_{II} \sim \varphi_{II}$ and $\cos\varphi_I/2 \sim 1$, then the *forward* current for

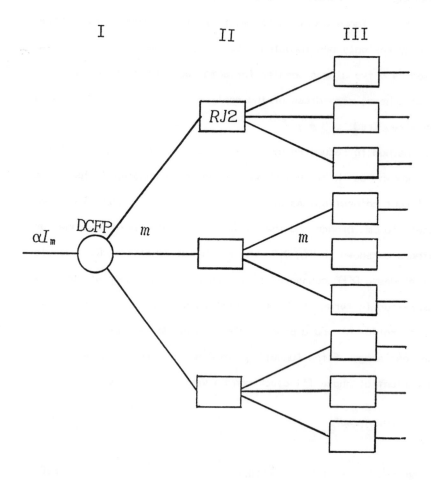

Fig.8.9 Coupling of RJ2 module by DCFP to be used for circuit fanout with better input and output isolation and more fanout capability.

the DCFP which is driven by clock phase I, shall be smaller or equal to the input signal therefore we choose a safety factor of $s=3$ in order to get the proper current flow in the fanout. That is,

$$\frac{\alpha I_m}{s} \geq \frac{m^2\alpha^2}{2} I_m \sin\varphi_{III} \tag{8.8}$$

$$m \leq \sqrt{\frac{2}{3\alpha \sin\varphi_{III}}} \tag{8.9}$$

This expression can further be improved by integrating the former hybrid transformer coupling scheme, which was shown in Fig.8.8, to the output side of RJ2. The hybrid transformer can cancel off the fanout noise mIF_1 if we choose the inductance of the hybrid circuit to be $M=L_J/2$ so that the current induced by the hybrid circuit is in reversing direction but with same amplitude as the fanout noise. However there is still a remainder of δIm noise which limits the number of permissible fanout as follows,

$$m \leq \frac{1}{\delta_2}\sqrt{\frac{2}{3\alpha\sin\varphi_{III}}} \tag{8.10}$$

Thus RJ2 can be used together with DCFP to reduce relay noise. One of the advantage of using RJ2 to work with DCFP is that RJ2 can alleviate the feedback logic loading of DCFP at one stage to the next stage if we place a RJ2 between two DCFPs. This is one of the example of application of IJL circuits.

8.4 Applications of DCFP to Memory and CPU Designs

Since DCFP operates on the superconducting temperature of $4.2°K$, it is desirable that at least the main memory also working in the same environment so that the access of information could be done with simpler interface and higher speed. As it is known that a JJ can be used to store a bit of information without necessary to provide the continuous support power<Harada85b>. However to read, write and address the information we need some controlled circuit. DCFP can be used to control the reading and writing of information to the memory. A memory unit using DCFP was proposed by E.Goto<Goto84c>and simulations of a 4x4 bits memory was also done<Harada85b> with satisfactory results. Memory cards will be packaged by the *antenna method*. By assuming packing density of $(50\mu m)^3$ per bit, a 10^9 bits MM can be packed in a $(5\ cm)^3$ cube surrounding the CPU which is placed in the centre with $(1\ cm)^3$ in size<Goto84d>. Cyclic Pipeline Machine was proposed to have the memory and CPU forming a closed loop pipeline connection in this setting.

In computer architecture pipeline is used as an economical means to increase the throughput of logical devices. Pipelining of memory with CPU is particularly well suited for the DCFP technology than silicon logic(ECL) for the following reasons. In silicon logic(ECL), insertion of a pipeline register increases the circuit delay time by several units of basic logic operations. Therefore a rather deep 10 to 20 stages of basic logic gates are placed between pipeline registers, so as to reduce the effect of the extra delay. In DCFP logic, since any DCFP functions as a latch, the pipeline scheme can be practiced without paying the penalty of extra delay time. The pipeline pitch time is equal to the clock period τ of the DCFP. Also, because of 3-phase clocking, 3 stages of (3-input) majority logic(with optional negations) can only be performed within a pipeline pitch time τ. Hence DCFP logic may be characterized as *High Pitch, Shallow Pipeline Logic*.

As mentioned earlier that MM can also be made from DCFP. If MM is made from DCFPs, it would be able to operate in a pipeline mode with the same pitch τ as the CPU. Hence MM will also be characterized by *High Pipeline Pitch*.

Cyclic Pipeline Architecture Machine was proposed based on this consideration. The basic features of this machine are that 40 processors and the MM are pipelining in order to achieve the maximum throughput of $10GIPS$. Each processor has its own memory partition within the MM and is accessed in every pitching time.

8.5 Conclusions

In this thesis we have studied a new kind of Josephson logic device known as a DCFP which is intended to be used for making very high speed computer. The strength of DCFP computer technology is that it is based on the well developed parametron computer technology as well as the very high switching speed Josephson junction. Furthermore the power consumption of this device is very small, it makes possible for higher integration without any serious heat sink problems.

The analysis we have done here shall help the potential user of this technology to understand and to utilize the device for various purposes. We have also proposed an economical and efficient software simulation tool for DCFP circuits design simulation. IJLs are studied as complement devices to DCFP. We have demonstrated how by using the software tool we can easily integrate IJL into DCFP study thus showing the way for the users to incorporate new Josephson devices into this design tool. Finally we discussed various interfacing problem when using DCFP for digital system design. DCFP not only can be used for processor design it also can be used for memory components, thus we have a complete DCFP computer technology.

With the pace of science and technology development, it is rather difficult to predict where the future computer technology will be lead to. What we can say about the future computer technology is largely based on the potential of basic research being done right now. Along this argument there are two important aspects which may affect the trend of computer technology. One of these is the device technology and the other is the computer architecture. Both have their limitations and problems yet they are not exclusive but complement to each other in some ways for our insatiable needs for computer power in the information society. We are optimistic on the DCFP computer technology standing on this arena for the following reasons: The DCFP device can attain even higher switching speed since the switching speed of DCFP depends on the switching speed of the Josephson junctions which is now improving by the discovery of new material and new method for junction fabrication. The junction is also making to consume less current and to operate on higher superconducting temperature with higher reliability. On the other hand the DCFP computer technology can be regarded as the extension of the conventional computer technology. Thus any progress on the conventional computer technology would definitely benefit to the DCFP computer technology. This is quite different from other more exotic computer technologies such as optical computer technology or molecular computer technology which deviates further from the conventional computer. Notwithstanding the future computer technology may be as diversified as the present ones, if not much more than now, to serve our different purposes of computer needs.

References

<Anderson63> P.W.Anderson and J.M.Rowell,Phys.Rev.Lett.10,230(1963).

<Bardeen57> J.Bardeen, L.N.Cooper and J.F.Schrieffer, Phys. Rev. 108,1175(1957).

<Faris80> S.M.Faris, W.H.Henkels, E.A.Valsamakis, H.H.Zappe, IBM J. of Res. Develop. 24, 143(1980).

<Fulton71> T.A.Fulton and R.C. Dynes, Solid-State Communications, 9,1069(1971).

<Gheewala79> T.Gheewala, IEEE J. Solid-State Circuit SC-14, 787(1979).

<Gheewala80a> T.R.Gheewala, IBM J. Res. Develop. 24,130(1980).

<Gheewala80b> T.R.Gheewala, IEEE Trans. Electr. Dev. ED-27, 1857(1980).

<Goto59> E.Goto, Proc.IRE,Vol47,p1309-1316,Aug.(1959).

<Goto60> E.Goto, K.Murata, K.Nakazawa, K.Nakagawa, T.Moto-oka, Matsuoka Y, Ishibashi Y, Ishda H, Soma T and Wada E. Proc. IRE Vol. EC-9 pp25-30, Mar(1960)

<Goto84a> E.Goto, p.48(In Japanese) Proc. of RIKEN Symposium on Josephson electronics,Mar(1984).

<Goto84b> E.Goto, p.84(In Japanese) Proc. of RIKEN Symposium on Josephson electronics,Mar(1984).

<Goto84c> E.Goto, p.96(In Japanese) Proc. of RIKEN Symposium on Josephson electronics,Mar(1984).

<Goto84d> E.Goto and K.Shimizu, Proc. of the 2nd Int. Symp. on Symbolic and Algebraic Computation by Computers, Wako-shi, p.17-1., Aug.(1984).

<Gueret80> P.Gueret, A.Moser, and P.Wolf, IBM J. Res. Develop. 24, 155(1980).

<Gueret77> P.Gueret, Th.O.Mohr, P.Wolf, IEEE Trans. Magn. MAG-13 52(1977).

<Harada85a> Y.Harada, N.Miyamoto, H.Nakane, U.Kawabe and E Goto, Proc. of RIKEN Second Sympm. on Josephson electronics, p.1(In Japanese), Mar(1985).

<Harada85b> Y.Harada, N.Miyamoto, U.Kawabe and E.Goto, Proc. of RIKEN Second Sympm. on Josephson electronics, p.9(In Japanese), Mar(1985).

<Henkels74> W.H.Henkels, IEEE Trans. Magnetics MAG-10, 860(1974).

<Henkels78> W.H.Henkels, H.H.Zappe, IEEE J.Solid-State Circuit SC-13, 591(1978).

<Henkels79> W.H.Henkels, J.Appl. Phys. 50, 8143(1979).

<IBM80> IBM Journal of Research and Development Vol.24,2.Mar(1980)

<Jaklevic65> R.C.Jaklevic, J.Lambe, J.E.Mercereau, and A.H.Silver, Phys. Rev. 140, A1628(1965).

<Josephson62> B.D.Josephson, Phys.Lett. 1,251(1962).

<Karman40> T.V.Karman and M Biot, Mathematical Methods in Engineering, p.98, McGraw-Hill, New York(1940).

<Klein78> M.Klein and D.J.Herrell, IEEE J.Solid-State Circuit SC. 13,577(1978).

<Likharev76> K.K.Likhara, G.M.Lapir and V.K.Semenov, Pis'ma v Zh. Tekhn. Fiz. (Sov. Techn. Phys. Lett.1, Vol.2, pp.809-814, Sep(1976); also U.S.S.R Patent No.539,333.

<Likharev77> K.K.Likharev, IEEE Trans. Magn., Vol.MAG-13, pp.242-244, Jan(1977).

<Loe84a> K.F.Loe and E.Goto, p.52, Proc. of RIKEN Symposium on Josephson electronics, Mar(1984).

<Loe84b> K.F.Loe, N.Osawa and E.Goto, The Second International Sympo-

sium on Symbolic and Algebraic Computation by Computer p.9-1, Aug(1984).

<Loe85> K.F.Loe and E.Goto, p.52, IEEE Trans, Magnetics Vol. MAG-21 No.2 Mar 85'. Proc. of RIKEN Second Symposium on Josephson electronics,Mar(1985).

<Martisoo67> J.Matisoo, Proc.IEEE(Letters)55,2052(1967). Also, Proc. IEEE 55, 172(1967).

<McCumber68> D.E.McCumber, Appl. Phys., 39,3113(1968).

<Moritsugu84> S.Moritsugu, N.Inada and E.Goto, The Second International Symposium on Symbolic and Algebraic Computation by Computer p.10-1, Aug(1984).

<Ohsawa84> N.Ohsawa, Josephson Circuit Simulation System Reference Manual, RIKEN Information Lab., (1984).

<Rowell63> J.M.Rowell, Phys.Rev. Lett.11,80(1963).

<Shapiro63> S.Shapiro, Phys. Rev. Lett. 11,80(1963).

<Shimizu84> K.Shimizu, E.Goto, Proc. of Riken Symp. on Josephson Electronics, pp.119-124 (in Japanese), Mar(1984).

<Shimizu85> K.Shimizu, E.Goto, Proc. of Riken Symp. on Josephson Electronics, pp.102 (in Japanese), Mar(1985).

<Stewart68> W.C.Stewart, Appl. Phys. Lett.12,277(1968).

<Takahasi60> H.Takahasi, Z.Kiyasu, Parametron, Parametron Institute, Sep(1960).

<VanDuzer81> T.Van Duzer, C.W.Turner, Principles of Superconductive Devices and Circuits, Elsevier, North Holland, Inc., New York(1981).

<Zappe73> H.H.Zappe, J.Appl. Phys., 44, 1371(1973).

<Zappe75a> H.H.Zappe, Appl. Phys. Lett. 27, 432(1975).

<Zappe75b> H.H.Zappe, IEEE J. Solid-State Circuits SC-10, 12(1975).

<Zappe75c> H.H.Zappe, US Patent Appl. No.341,002 , Jan(1975).

<Zappe77> H.H.Zappe, IEEE Trans. Magn., MAG-13,41(1977).

<Zappe83> H.H.Zappe, Compcon Spring 83(1983).

APPENDIX

APPENDIX

The Parametron, a Digital Computing Element which Utilizes Parametric Oscillation*

EIICHI GOTO†

Summary—The following is a brief description of the basic principles and applications of the parametron, which is a digital computer element invented by the author in 1954. A parametron element is essentially a resonant circuit with a nonlinear reactive element which oscillates at one-half the driving frequency. The oscillation is used to represent a binary digit by the choice between two stationary phases π radians apart. The basic principle of logical circuits using the parametron is explained, and research on and applications of parametrons in Japan are described.

I. INTRODUCTION

IN keeping with the remarkable progress of electronic computers in recent years, studies on digital computing elements and memory devices have been energetically conducted in various laboratories. Among them, one will find new applications of physical phenomena and effects that have never before been utilized in the field of electronics; the cryotron, which uses superconductivity, and the spin echo memory are typical examples.

In 1954 the author discovered that a phenomenon called parametric oscillation, which had been known for many years, can be utilized to perform logical operations and memory functions, and gave the name "Parametron" to the new digital component made on this principle [1], [21]–[23].

A digital computing circuit made of parametrons may consist only of capacitors, ferrite-core coils and resistors, while diodes and rectifiers may be dispensed with. The parametron, therefore, is considered to be extremely sturdy, stable, durable, and inexpensive. Owing to these advantages, intensive studies have started in several laboratories in Japan to apply parametrons to various digital systems. At present, nearly half of the Japanese electronic computers in operation use parametrons for logical elements. Further applications have been made to such devices as telegraphic equipments, telephone switching systems and numerical control of machine tools.

Parametric oscillation, from which the name "Parametron" derives, is not a unfamiliar phenomenon—a playground swing and Melde's experiment are examples of parametric oscillations in mechanical systems.

In order to drive a swing, the rider bends and then straightens his body and thereby changes the length l between the center of gravity of his body and the fulcrum of the ropes. The swing is a mechanical resonant system and its resonant frequency is determined by this length l and the gravitational constant g. The oscillation of the swing is energized by the periodic variation of the parameter l which determines the resonant frequency. Similarly, in Melde's experiment, shown in Fig. 1, a periodic variation is given to the tension, which is a parameter that determines the resonant frequency of the string. In this case, the exciting energy which varies the tension is supplied from a tuning fork of resonant frequency $2f$, which is twice the resonant frequency f of the string. In other words, the oscillating frequency of the string is a subharmonic equal to half the frequency of the energy source, that is, it is the second subharmonic. The mechanism of building up of this subharmonic is shown in Fig. 2. As the string moves away from the equilibrium position, the tension is weakened and the maximum amplitude increases; as the string returns to the center position, the tension is strengthened and the kinetic energy increases.

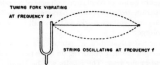

Fig. 1—Melde's experiment.

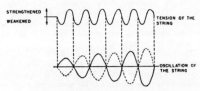

Fig. 2—Build up of oscillation of the string.

In an electrical system, inductance and capacitance are the parameters which determine the resonant frequency. Parametric oscillation therefore can be produced in a resonant circuit by periodically varying one of the reactive elements composing the resonant circuit [18].

A parametron element is essentially a resonant circuit with a reactive element varying periodically at frequency $2f$ which generates a parametric oscillation at the subharmonic frequency f. In practice, the periodic

---

* Original manuscript received by the IRE, December 9, 1958; revised manuscript received, May 14, 1959.
† Dept. of Physics, University of Tokyo, Tokyo, Japan.

variation is accomplished by applying an exciting current of frequency $2f$ to a balanced pair of nonlinear reactors, such as ferrite-core coils and nonlinear capacitors made of ferroelectric material or of the barrier capacitance of semiconductor junctions.

The subharmonic parametric oscillation thus generated has a remarkable property in that the oscillation will be stable in either of two phases which differ by π radians with respect to each other. Utilizing this fact, a parametron represents and stores one binary digit, "0" or "1," by the choice between these two phases, 0 or π radians. The solid line and the dotted line in Fig. 2 illustrate the building up of these two kinds of oscillation.

Under certain resonance conditions, the oscillation generated in the parametron is "soft," that is, it is easily self-started from any small initial amplitude. In this case, the choice between the two stable phases of the oscillation having a large amplitude can be made by controlling the phases of the small initial oscillation. This fact may be regarded as amplification and its mechanism may best be understood as superregeneration with the phase of the oscillation quantized to two states. In order to make use of this effectively, quenching means are provided in parametron circuits to interrupt parametric oscillation. Besides the memory and amplifying action, parametrons can also perform various logical operations based on a majority principle by applying the algebraic sum of oscillation voltages of an odd number of parametrons to another parametron in which the algebraic sum voltage works as the small initial oscillation voltage.

Mathematical studies on parametric oscillations of small amplitude in a linear region have been conducted in detail in the past. The results will be found in textbooks on differential equations under such headings as linear differential equations with periodic coefficients, Mathieu's equation, Hill's equation, and Floque's theorem [16], [17]. However, in order to describe the actual behavior of parametrons quantitatively, one has to take nonlinearity into consideration, and this will be treated in the Appendix.

The application of parametric oscillation to amplifying electrical signals is not a new idea. We find in Peterson's patent of 1932 [29], an idea for an amplifier based on the same principle as the parametric amplifier, which is now one of the most discussed topics in the field of electronics. In a parametric amplifier, two resonant circuits, respectively tuned to signal frequency f_s and idling frequency f_i, are coupled together regeneratively through a nonlinear reactor to which is applied a voltage of pumping frequency f_p, satisfying the condition $f_p = f_i + f_s$. A parametric amplifier performs regenerative amplification of signals and may produce, as well, a pair of spontaneous oscillations at frequency f_s and f_i.

A parametron producing a subharmonic oscillation may be regarded as a degenerative case of a parametric amplifier, in which the two resonant circuits for f_s and f_i are reduced to a single common circuit, so that $f_s = f_i = f$, and $f_p = 2f$. Consequently, the basic principle of the amplifying mechanism of the parametron may be considered the same as that of the parametric amplifier. The degeneracy in the number of resonant circuits, however, makes possible the phase quantizing nature of the oscillation. While this is generally unfavorable for amplifying ordinary continuous waves, it is very useful for representing and storing a binary digit in the parametron.

Parametric oscillation of the second subharmonic mode in an electrical system has been known for many years and has been applied to frequency dividers [18]. On the other hand, the idea that two stable phases exist and can be applied to digital operations can be found only in a patent [30] of the late Professor von Neumann, so far as the author knows. Von Neumann proposed, completely independent of the author, a scheme similar to the parametron. His idea, however, seems to have not yet been developed into practical use.

If the resonance condition of a circuit which produces the subharmonic parametric oscillation is slightly altered, a "hard" oscillation, i.e., not self-starting, will be produced. This circuit, generally, has three stable states, namely, "no oscillation," "oscillating at 0 phase," and "oscillating at π-radian phase." In Japan such an element is usually called a "tristable parametron," while in the case of "soft" oscillation it is called a "bistable parametron." In the hard oscillation circuit, i.e., the tristable parametron, a binary digit can be represented by the presence or absence of oscillation. This scheme has also been proposed independently by Clary [31].

II. Basic Principle

The parametron is essentially a resonant circuit in which either the inductance or the capacitance is made to vary periodically. Fig. 3 shows circuit diagrams for parametron elements. The parametron element in Fig. 3(a) consists of coils wound around two magnetic ferrite toroidal cores $F1$ and $F2$, a capacitor C, and a damping resistor R, and a small toroidal transformer T. Each of the cores $F1$ and $F2$ has two windings and these are connected together in a balanced configuration, one winding $L = L' + L''$ forming a resonant circuit with the capacitor C and being tuned to frequency f. An exciting current, which is a superposition of a dc bias and a radio frequency current of frequency $2f$, is applied to the other winding, $l' + l''$, causing periodic variation in the inductance $L = L' + L''$ of the resonant circuit at frequency $2f$.

The parametron in Fig. 3(b) consists of two nonlinear capacitors C' and C'' which form the resonant circuit with the inductance L. An exciting voltage of frequency $2f$ is supplied between the neutral point of the two nonlinear capacitors C', C'' and the neutral point of the inductance L, causing periodic variation in the tuning capacitance $C(1/C = 1/C' + 1/C'')$ at frequency $2f$. As

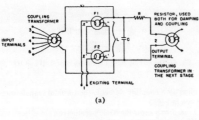

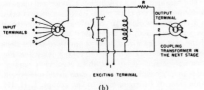

Fig. 3—Circuit diagram of parametrons, (a) Magnetic type. (b) Capacitive type.

the results are entirely analogous in both cases, the following explanations will be given only for the former case.

The operation of the parametron is based on a spontaneous generation of a second-subharmonic parametric oscillation, that is a self-starting oscillation of frequency f, in the resonant circuit. Parametric oscillation is usually treated and explained in terms of Matheiu's equation. A more intuitive explanation, however, may be obtained by the following consideration.

Let the inductance L of the resonant circuit be varied as

$$L = L_0(1 + 2\Gamma \sin 2\omega t) \quad (1)$$

where $\omega = 2\pi f$, and let us assume the presence of a sinusoidal ac current I_f in the resonant circuit at frequency f, which can be broken down into two components as follows:

$$I_f = I_s \sin(\omega t) + I_c \cos(\omega t). \quad (2)$$

Then, assuming that the rate of the variation of amplitudes of the sine and cosine components, I_s and I_c, are small compared with ω, the induced voltage V will be given by

$$V = d/dt(LI_f) = \omega L_0(I_s \cos \omega t - I_c \sin \omega t)$$
$$+ 3\Gamma\omega L_0(I_s \sin 3\omega t + I_c \cos 3\omega t)$$
$$+ \Gamma\omega L_0(-I_s \sin \omega t + I_c \cos \omega t). \quad (3)$$

The first term shows the voltage due to a constant inductance L_0, and the second term or the third harmonic term may be neglected in our approximation, since it is off resonance. The third term, which is essential for the generation of the second subharmonic, shows that the variable part of the inductance behaves like a negative resistance $-r = -\Gamma\omega L_0$ for the sine component I_s, but behaves like a positive resistance $+r = \Gamma\omega L_0$ for the cosine component I_c.

Therefore, provided that the circuit [Fig. 3(a)] is nearly tuned to f, the sine component I_s of any small oscillation (Ⓐ in Fig. 4), will build up exponentially (Ⓑ in Fig. 4), while its cosine component will damp out rapidly. If the circuit were exactly linear, the amplitude would continue to grow indefinitely. Actually, the nonlinear B-H curve of the cores causes detuning of the resonance circuit and hysteresis loss also increases with increasing amplitude, so that a stationary state (Ⓒ in Fig. 4) will rapidly be established, as in vacuum-tube oscillators. Details of the amplitude limiting mechanism, which is essentially a nonlinear problem, will be treated in the Appendix. The solution of the problem will be illustrated most intuitively by showing the locus of the sine and cosine components, I_s and I_c in the (I_s, I_c) plane. Fig. 5 shows an example of such loci for a typical case $\alpha = 0$, $\delta = \Gamma/2$ (cf. Appendix). The abscissa represents the sine component I_s and the ordinate, the cosine component I_c. If we introduce polar coordinates (R, ϕ) in the (I_s, I_c) plane, it will be seen easily from (1) that R and ϕ, respectively, indicate the instantaneous amplitude and phase of the oscillation. The saddle point at the origin indicates the exponential build up of oscillation which is in a definite phase relation to the excitation wave of frequency $2f$. Spiral points A and A' in the figure indicate the stable states of stationary oscillation. The existence of two possible phases in this oscillation which differ by π radians from each other, corresponding to A and A', should be noted. These two modes of oscillation are respectively shown by the solid line and dotted line in Fig. 4. An especially important feature is that the choice between these two modes of stationary oscillation is effected entirely by the sign of the sine component of the small initial oscillations that have existed in the circuit (A in Fig. 4). In other words, the choice between A and A' in Fig. 5 depends on which side of the thick curve B-B' (called separatrix) the point representing the initial state lies. An initial oscillation of quite small amplitude is sufficient to control the mode or the phase of stationary oscillation of large amplitude which is to be used as the output signal. Hence, the parametron has an amplifying action which may be understood as superregeneration. The upper limit of this superregenerative amplification is believed to be determined only by the inherent noise, and an amplification of as high as 100 db has been reported.[1]

The existence of dual mode of stationary oscillation can be made use of to represent a binary digit, "0" and "1" in a digital system, and thus a parametron can store

---

[1] A personal communication from Z. Kiyasu, of Electrical Communication Laboratory, Nippon Telephone and Telegraph Co., Tokyo, Japan.

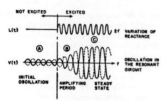

Fig. 4—Oscillation of parametrons.

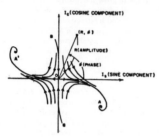

Fig. 5—The amplitude-to-phase (R, ϕ) locus of an oscillating parametron.

Fig. 6—The exciting current of three groups, I, II and III.

1 bit of information. However, oscillation of parametrons in this stationary state is extremely stable, and if one should try to change the state of an oscillating parametron from one mode to another just by directly applying a control voltage to the resonant circuit, a signal source as powerful as the parametron itself would be necessary. This difficulty can be got around by providing a means for quenching the oscillation, and making the choice between the two modes, *i.e.*, the rewriting of information, by a weak control voltage applied at the beginning of each building up period, making use of the superregenerative action.

Actually, this is done by modulating the exciting wave by a periodic square wave which also serves as the clock pulse of the computer. Hence, for each parametron there is an alternation of active and passive periods, corresponding to the switching on and off of the exciting current. Usually, the parametron device uses three clock waves, labeled I, II and III, all having the same pulse recurrence frequency, which are switched on and off one after another in a cyclic manner as shown in Fig. 6. This method of exciting each of the parametrons in a digital system with either one of the three exciting waves I, II and III is usually called the "three beat" or the "three subclock" excitation.

III. Basic Digital Operations by Parametrons

Digital systems can be constructed using parametrons by intercoupling parametron elements in different groups by a coupling element, the toroidal transformers shown in Fig. 3.

Figs. 7 to 9 show the basic parametron circuits.

The parametron is a synchronous device and operates in rhythm with the clock pulse. Each parametron takes in a new binary digit ("1" or "0") at the beginning of every active period, and transmits it to the parametrons of the next stage with a delay of one-third of the clock period. This delay can be used to form a delay line. Fig. 7 shows one such delay line which consists of parametrons simply coupled in a chain, each successive parametron element belonging each to the groups, I, II, III, I, · · · . Hence, the phase of oscillation of a parametron in the succeeding stage will be controlled by that in the preceding stage, and a binary signal x applied to the leftmost parametron will be transmitted along the chain rightwards in synchronism with the switching of the exciting currents. Hence, the circuit may be used as a delay line or a dynamic memory circuit.

Fig. 8 shows how logical operations can be performed using parametrons. In the figure, the outputs of the three parametrons X, Y and Z in stage I, which are all oscillating at a voltage V, are coupled to the single parametron U in stage II with a coupling factor k. As the effective phase control signal acting on U is the algebraic sum of the three signals from X, Y and Z, each of which has the value $+kV$ or $-kV$, the mode of U representing a binary signal u will be determined according to the majority of the three binary signals x, y and z, respectively represented by the oscillation modes of X, Y and Z. It would be possible, in principle, to generalize the majority circuit of Fig. 8(a) to 5, 7, 9, · · · inputs, that is, to any odd number of inputs. In practice, however, the nonuniformity in the characteristics of each parametron causes disparity in the input signals and makes the majority decision inaccurate, and this fact limits the allowable number of inputs to 3 or 5 in most cases.

It is easily seen that the majority operation just outlined includes the basic logical operations "and" and "or" as special cases. Suppose that one of the three inputs in Fig. 8(a), say z, is fixed to a constant value "1," then we obtain a biased majority decision on the remaining two inputs x and y, and the resulting circuit gives "x or y" as shown in Fig. 8(c). Similarly, if z is fixed to a constant value "0," we obtain a circuit for "x and y" as shown in Fig. 8(d). These constant signals are actually derived from a special parametron called constant parametron, or some other voltage source equiva-

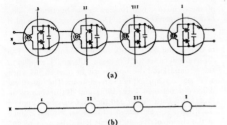

Fig. 7—A parametron delay-line circuit.

lent to it. If 2 out of 5 inputs in a "5-input majority operation" are made constants, either "1" or "0," we shall obtain either a "or" or "and" circuit respectively for three input variables x, y and z, as shown in Figs. 8(e) and 8(f).

"Complementation" or "not" operation can be made most simply in parametron circuits. In order to change the binary signal "1" into "0" and vice versa, we have only to reverse the polarity of the input signal, and this can be done by coupling two parametrons in reversed polarity as shown in Fig. 9. In the schematic diagram, such coupling in reversed polarity will be indicated by a short bar in the coupling line as shown in Fig. 9.

Since a digital system of any complexity can be synthesized by combining the four basic circuit elements, namely "delay," "and," "or," and "not," it will be seen that a complete digital system, e.g., a general purpose electronic computer, can in principle be constructed using only one kind of circuit element—the parametron. It should be noted that the above conclusion presupposes that some means for logical branching, that is amplification of signal power, is provided. Now parametrons have a large superregenerative amplification and the output of a single parametron can supply input signals to a rather large number of parametrons in the next stage. For the parametrons currently used, the maximum allowable number of the branching is from 10 to 20. This feature adds flexibility in design of digital systems using parametrons.

IV. Simple Examples of Parametron Circuits

A complete digital apparatus may consist of hundreds or thousands of parametrons, coupled to each other by wires (via resistors and transformers) to form a network. Such a network of parametrons may be conveniently described by a schematic diagram.

At this point we will give a short summary of the rules and conventions for schematic diagrams currently in use in Japan.

Each parametron is represented by a small circle, as shown in the figures. Each pair of circles is connected by a line if corresponding parametrons are coupled, one

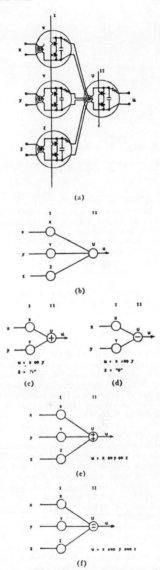

Fig. 8—A three-input majority operation circuit. (a) Circuit diagram. (b) The schematic diagram of (a). (c) An "or" circuit for two inputs. (d) An "and" circuit for two inputs. (e) An "or" circuit for three inputs. (f) An "and" circuit for three inputs.

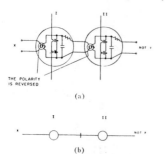

Fig. 9—(a) A "not" circuit. (b) The schematic diagram of the "not" circuit (a).

line being used per unit coupling intensity. Hence, a double line between circles will indicate that both parametrons are coupled at double coupling intensity (cf. Fig. 14). A short bar across any coupling line denotes complementation, that is, both parametrons are coupled with reversed polarity (Fig. 9), and otherwise it is understood that they are coupled in the same polarity.

If not specified, parametrons are supposed to be excited with the three-beat excitation. Accordingly, only parametrons belonging to different groups (I, II and III) can be coupled, and the information is transmitted along these lines always in the direction: I→II, II→III and III→I. Therefore, each coupling line has a definite direction of transmission, and to show this direction, usually the output lines from a parametron will come from the right side of the circle and go into the left side of another circle as input to it. As has been explained in Section III, there may be many parametrons which take some of their input signals from special parametrons called constant parametrons. These belong to a special triplet of parametrons, connected in a ring and always holding the digit "1," and serving as the phase reference. Since there may be a great many lines that come from these constant parametrons, these lines are usually omitted from the diagram, in order to avoid complication, and one "+" symbol is inscribed in the circle per unit constant input of positive polarity, and also one "−" symbol is inscribed per unit constant input of negative polarity. Accordingly, a circle with a "+" having two input lines corresponds to an "or" element, a circle with a "−" having two input lines corresponds to an "and" element, and a circle with "++" having three input lines corresponds to a 3-input "or" element, etc. It should be noted that the distinction between "0" and "1" in a parametron circuit is only possible by referring to the oscillation phase of these constant parametrons, since the phase is a relative concept.

The following figures show some simple examples of actual parametron circuits in schematic diagrams. The reader will not find it difficult to trace the functioning of these circuits. Fig. 10 shows a parametron flip-flop or a 1-bit memory circuit. Three parametrons, coupled in ring form, are required to store 1 bit of information. In Figs. 10(a) and 10(b) it is assumed that the signals in the set and reset inputs are both normally "0." The flip-flop will be set to "1" when a "1" signal is applied to the set input, and the flip-flop will be reset to "0" when a "1" signal is applied to the reset input. The functional difference between Figs. 10(a) and 10(b) consists in that, when both the set and the reset signal are applied simultaneously, the stored information will not change in the circuit of Fig. 10(a), but it will be reset to "0" in the circuit of Fig. 10(b).

Fig. 10(c) shows a flip-flop with a gate. As long as "0" is applied to the gate, the stored information does not change, but when "1" is applied to the gate, the signal from the input is transferred to the flip-flop.

Fig. 11 shows three stages of binary counting circuits connected in cascade, thus forming a scale-of-8 counter. Three flip-flops are included in this circuit to store a 3-bit count. In the quiescent state, in which "0" is applied to the input, the bits stored in each flip-flop do not change, but each time a "1" is applied to the input for a single clock period, the registered binary number is increased by 1 (mod 8). Figs. 12, 13 and 14 show respectively a binary full-adder circuit for three input signals, a parity check circuit for five input signals and a circuit for "x and (y or z)." These examples will show how majority operations can be made use of advantageously compared to the "and" and "or" operations. These circuits would have required many more parametrons if they were composed of "and" and "or" operations as in the usual diode networks. Flexibility of circuit design by use of a three- or five-input majority operation will be regarded as one of the characteristic features of parametron circuits.

The reason for the necessity of three subclock waves I, II and III, shown in Fig. 4, will be shown in Fig. 15. In Fig. 15, $P1$, $P2$, etc., indicate parametrons and $C1$, $C2$, etc., indicate coupling elements provided between parametrons. Each of the parametrons is supposed to be excited with either one of the two kinds of radio-frequency waves, I', and II', as shown in Fig. 16. These two waves are switched on and off alternately and will be called the two subclock exciting waves. If the coupling between two parametrons consists of a passive linear circuit, which is essentially a bilateral system, and if the two parametrons $P1$ and $P3$ in Fig. 15 are generating oscillations, voltages will be transmitted to $P2$ from both $P1$ and $P3$ with substantially the same intensity, and the phase-controlling action of $P2$ will become uncertain. Therefore, in order to use the two subclock exciting waves I', and II' in a parametron circuit, it is necessary to use unilateral coupling means. This may be accomplished by using a unilateral element, such as vacuum tubes and transistors, or by varying the coupling coefficient of the coupling elements as $K1$ and $K2$ in Fig. 16 by means of applying a gating signal to

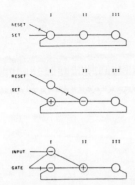

Fig. 10—Flip-flop circuits. (a) A flip-flop circuit. (b) A flip-flop circuit. (c) A flip-flop circuit with an input gate.

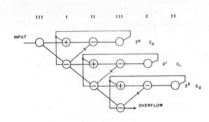

Fig. 11—Three stages of binary counters forming a scale of eight circuits.

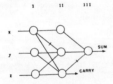

Fig. 12—A binary full-adder circuit.

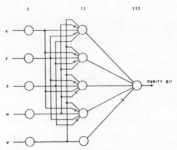

Fig. 13—A parity-check circuit for five input signals.

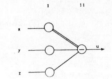

Fig. 14—A circuit for "x and (y or z)."

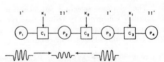

Fig. 15—Coupling system for two-subclock excitation.

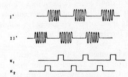

Fig. 16—Two-subclock excitation.

nonlinear elements, such as diodes and magnetic cores [27].

In the three-beat or the three-subclock exciting method, each of the parametrons will be excited once in every clock cycle at a definite time. In this respect the method may be called stationary excitation. On the other hand, we may think of a more general method, usually called "non-stationary excitation" or "gated excitation" in Japan, in which the excitation of parametrons is switched in accordance with gating signals [28]. Fig. 17 shows a selecting circuit using the "gated excitation." $S_1, S_2, \cdots, S_n$ indicate binary phased information sources and $P_1, P_2, \cdots, P_n$ are gating parametrons. Supposing that the exciting wave I is applied selectively to only one of the gating parametrons, say P_2, by controlling the excitation with a gating signal so as to produce oscillation only in P_2, the information from S_2 will be selectively transmitted to the parametron P, since the oscillation of P_2 is controlled by S_2 and the oscillation of P is controlled by that of P_2. For comparison, Fig. 18 shows a selecting circuit for one out of four channels using the three-subclock (stationary) ex-

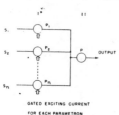

Fig. 17—Channel-selecting circuit using gated excitation.

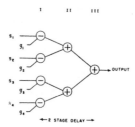

Fig. 18—Channel-selecting circuit using three-beat excitation.

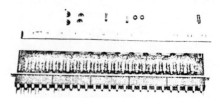

Fig. 19—A parametron unit (25 parametrons).

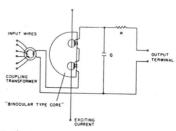

Fig. 20—The circuit of a parametron with a "binocular type core" and a series type coupling transformer.

citation. The channel selecting arrangement using gated excitation will generally reduce the access time and the number of parametrons at the expense of employing rather complicated exciting circuits.

V. CHARACTERISTIC FEATURES OF PARAMETRONS

Fig. 19 shows a commercial unit composed of 25 parametrons and the component parts. In this unit a ferrite disc with two small holes (known as a "binocular type core" [7]) is used instead of the two toroids in Fig. 3. The coupling transformer consists of a single-turn coil wound on a ferrite toroid and is connected in series to the resonant circuit as shown in Fig. 20. As the life of parametrons is considered to be practically permanent, the parametron units are usually not made in a "plug-in" style, but are directly wired into the logical networks.

As may be seen from Figs. 3, 7, and 8, the wiring of parametron circuits is done in an unusual way. A wire connected to the output terminal of a parametron in a preceding stage is passed through the coupling transformers of all the parametrons in the succeeding stage which are to receive the input signal from the parametron in the preceding stage. This has resulted in a remarkable simplification in the construction of complex logical networks, such as general purpose computers, since the whole system can be assembled from identical standardized units, and the units can be wired to form the specific machine using only wires and with a minimum number of soldering points. Table I shows the typical characteristics of commercial parametrons in Japan.

For application to digital computers we are most concerned in the speed of operation, which is essentially determined by the clock frequency F_k. The upper limit of F_k is limited by the rates of building up and damping of parametric oscillation.

From (3) it follows that the oscillation builds up proportionally to $e^{\pi \Gamma f t}$ (cf. Appendix), and hence the maximum clock frequency will be proportional to the product Γf, which we call the "figure of merit" of the parametron, owing to its analogy to the figure of merit of vacuum tubes $g_m/2\pi C$.

TABLE I
CHARACTERISTICS OF COMMERCIAL PARAMETRON UNITS

	High Speed Type	Standard Type	Low Power Type
Exciting frequency $2f$	6 mc	2 mc	200 kc
Maximum clock frequency	140 kc	25 kc	2 kc
Exciting power per one parametron for continuous excitation	120 mw	30 mw	5 mw
DC bias	0.6 amp	0.6 amp	0.6 amp
Maximum number of inputs	3 or 5	3 or 5	3 or 5
Maximum number of output branching	12	15	15
Coupling coefficient*	−35 db	−40 db	−40 db

* *Note:* The coupling coefficient k is defined as the ratio:

$$k = \frac{\text{voltage of unit input measured at the resonant circuit}}{\text{voltage of stationary oscillation}}$$

The figure of merit of a parametron naturally depends on the frequency and amplitude of the exciting current, and the value for conventional parametrons in normal operation lies between 20kc and 1.5 mc. For reliable operation, the clock frequency should be chosen around $Pf/10$ if the coupling factor is $k = 1/50$ (-34 db), and hence the upper limit for the clock frequency is about 150 kc for commercial parametron units.

In the past, most effort has been made to develop parametrons using variable inductance, but it is apparent that the same principle applies when the capacitor is the variable element. Parametrons using ceramic nonlinear capacitors (barium titanate) have been studied by Oshima and Kiyasu [8].

Studies of parametrons using the variable barrier capacitance of germanium and selenium diodes have also been made, and a parametric oscillation at $f = 60$ mc has been realized [5], [9].

and will limit their use in small-scale digital devices. At the present stage, parametrons seem to be unfavorably compared with vacuum tubes and transistors in speed of operation, but this point may be much improved by further development.

VI. APPLICATION

All the characteristics of parametrons just mentioned make them ideally suited to applications in large-scale digital devices, and particularly to general purpose digital computers. Soon after the invention of parametrons in 1954, a project was launched to construct general purpose computers using control and arithmetic units entirely composed of parametrons. At present, nearly half of the digital electronic computers built in Japan are parametron computers [11], [14]. Table II shows the characteristics of these computers.

TABLE II
THE CHARACTERISTICS OF GENERAL PURPOSE PARAMETRON COMPUTERS

Type (Date of Completion)	Place of Installation	The Number of Parametrons (Number System)	Exciting Frequency	Clock Frequency	Speed of Operation (for Fixed Point) Including Access		Main Memory	Power
					Addition	Multiplication		
FACOM 212 (March, 1959)	Fuji Elec. Co. Kawasaki	8000 (Decimal)	2 mc	10 kc	4 ms	15 ms	49 words Core Matrix	5 kw
HIPAC-1 (December, 1957)	Cint. Lab. Hitachi Elec. Co. Kokubunji, Tokyo	4400 (Binary)	2 mc	10 kc	10 ms	19 ms	1024 words Magnet Drum	6 kw
MUSASINO-1 (March, 1957)	Elec. Communication Lab. Musasino, Tokyo	5400 (Binary)	2 mc	6 kc	4 ms	20 ms	256 words Core Matrix	5 kw
NEAC-1101 (April, 1958)	Cent. Lab. Nippon Elec. Co. Kawasaki	3600 (Binary)	2 mc	20 kc	3.5 ms	8 ms	128 words Core Matrix	5 kw
PC-1 (March, 1958)	Department of Physics University of Tokyo Tokyo	4200 (Binary)	2 mc	15 kc	270 µs	3.4 ms	256 words Core Matrix	3 kw
PC-2* (August, 1959)		9600 (Binary)	6 mc	100 kc	40 µs	340 µs	1024 words Core Matrix	10 kw
SENAC-1 (November, 1958)	Elec. Communication Lab. University of Tohoku Sendai	9600 (Binary)	2 mc	20 kc	2 ms	3 ms	160 words Magnetic Drum	15 kw

* *Note:* The construction of PC-2 will be completed in August 1959.

Parametrons are composed of capacitors, resistors and coils with ferromagnetic cores which are all stable and durable components. Unlike the more conventional switching circuits using magnetic amplifiers, parametrons require no diodes for their operation. These features guarantee for parametron circuits extremely high reliability and long life. In several digital computers now in operation in Japan, troubles with parametrons are extremely rare.

The necessity of a high-frequency power supply may be one of the inherent disadvantages of parametrons

In the core matrix memory of these parametron computers, an entirely new method, proposed by the author in 1955 [24]–[26], is employed both for reading and writing. Writing is effected by impressing on each memory core the superposition of two ac currents, supplied from parametrons and having frequencies of f and $f/2$. Reading is also effected with parametrons by amplifying and sensing the phase of the second harmonic component of frequency f which is generated in each memory core by impressing an ac current of frequency $f/2$ on it.

The new method is called "dual frequency memory system," and the following are considered to be characteristic features:

1) Memory cores are driven by output ac currents of parametrons.
2) Only two windings, X and Y, pass through each memory core.
3) Reading is nondestructive.

The details will be discussed in a separate paper to follow.

The application of parametrons to other digital devices has also been made in a number of laboratories. The Japan Overseas Telephone and Telegraph Company has constructed regenerative repeaters, telegraph code converters which convert Morse code to five-unit teleprinter code [6], and ARQ (automatic request) systems, which have all been in commercial use for some years.

The Japan Telegraph and Telephone Corporation has built a number of experimental common-control telephone switching systems, employing parametrons in control circuits [15]. The Fuji Electric Company and the Government Mechanical Laboratory have built experimental numerically controlled machine tools [13], in which parametrons are used for all numerical and control operations. Among other applications are automatic recording systems for a meson monitor used in cosmic-ray observation and multichannel pulse-height analyzers for nuclear research [4].

Appendix
Amplitude Limiting Mechanism of the Parametron

First, we shall derive the equation governing the oscillation in a parametric resonant circuit including a variable inductance $L(t)$ as shown in Fig. 21.

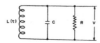

Fig. 21—A parametrically-excited resonant circuit.

The voltage V in the resonant circuit will be given by

$$V = \frac{d}{dt}(Li) \qquad (4)$$

where i is the current passing through the inductance. From Kirchhoff's law, we obtain

$$i + \frac{V}{R} + \frac{d}{dt}(CV) = 0. \qquad (5)$$

We shall assume that the inductance is varying as

$$L(t) = L_0(1 + 2\Gamma \sin 2\omega t). \qquad (6)$$

Putting

$$I = \frac{L}{L_0} i = (1 + 2\Gamma \sin 2\omega t)i \qquad (7)$$

$$\delta = \frac{1}{\omega CR} = \frac{1}{Q} \qquad (8)$$

$$\frac{1}{CL_0} = \omega^2(1 + \alpha) \qquad (9)$$

and assuming that Γ and α are much smaller than unity so that $(1+\alpha)(1+2\Gamma \sin 2\omega t)^{-1}$ may be replaced by $1+\alpha-2\Gamma \sin 2\omega t$, (6) will be rewritten as

$$\left[\frac{d^2}{dt^2} + \delta\omega \frac{d}{dt} + \omega^2(1 + \alpha - 2\Gamma \sin 2\omega t)\right]I = 0. \qquad (10)$$

We may call δ the loss factor, α the detuning of the resonant circuit from the second-subharmonic frequency ω, Γ the modulation index of the inductance and L_0 the constant part of the inductance. As the difference between I defined by (7) and the actual current i is of the order of Γ, the results to be obtained from the following analysis of I may be regarded as substantially the same as that of the actual current i when Γ is small.

In case α and δ are constants, (10) represents a linear differential equation, well known as Mathieu's equation [16]-[18]. In practice, however, ferromagnetic cores are used in the inductance to effect the variation, and with increasing amplitude the nonlinear B-H curve will cause detuning of the resonant circuit and hysteresis loss will also increase. Consequently, the loss δ and the detuning α of (10) will generally be functions of the amplitude I^2 and (10) becomes a nonlinear differential equation.

Now, we shall assume the presence of nonlinearity of the form βI^2 as the detuning. Then (10) becomes

$$\left[\frac{d^2}{dt^2} + \delta\omega \frac{d}{dt} + \omega^2(1 + \alpha + \beta I^2 - 2\Gamma \sin 2\omega t)\right]I = 0. \qquad (11)$$

Breaking down I into two sinusoidal components as

$$I = I_s \sin \omega t + I_c \cos \omega t, \qquad (12)$$

(11) will be rewritten as

$$[2\dot{I}_s\omega + \delta\omega^2 I_s + \alpha\omega^2 I_c - \Gamma\omega^2 I_c + \tfrac{3}{4}\beta\omega^2(I_s^2 + I_c^2)I_c] \cos \omega t$$
$$+ [-2\dot{I}_c\omega - \delta\omega^2 I_c + \alpha\omega^2 I_s - \Gamma\omega^2 I_s + \tfrac{3}{4}\beta\omega^2(I_s^2 + I_c^2)I_s] \sin \omega t$$
$$+ [\ddot{I}_c + \delta\omega \dot{I}_c] \cos \omega t$$
$$+ [\ddot{I}_s + \delta\omega \dot{I}_s] \sin \omega t$$
$$+ [\Gamma\omega^2 I_s + \tfrac{1}{4}\beta\omega^2 I_c^3 - \tfrac{3}{4}I_s^2 I_c] \cos 3\omega t$$
$$+ [-\Gamma\omega^2 I_c - \tfrac{1}{4}\beta\omega^2 I_s^3 + \tfrac{3}{4}I_s I_c^2] \sin 3\omega t = 0. \qquad (13)$$

In order to obtain an approximate solution of the nonlinear differential equations (11) or (13), we shall assume that α, Γ and δ are much smaller than unity.

Then I_s and I_c in (13) will vary much more slowly than ω, and the third and fourth terms of (13) may be neglected since they are much smaller than ω^2. The third harmonic terms may also be neglected since they are off resonance and thus we will obtain the following approximate equations for I_s and I_c:

$$\frac{2}{\omega} \dot{I}_s = -\delta I_s + \Gamma I_c - (\alpha + \tfrac{3}{4}\beta(I_s^2 + I_c^2))I_c$$

$$\frac{2}{\omega} \dot{I}_c = -\delta I_c - \Gamma I_s + (\alpha + \tfrac{3}{4}\beta(I_s^2 + I_c^2))I_s. \quad (14)$$

Each term of (14) has the following intuitive meaning: the first term with δ represents the loss in the circuit; the second term with Γ indicates negative resistance effect for the sine component I_s and damping (positive resistance) effect for the cosine component I_c; and the third term with α and β represents detuning of the resonant circuit which is a function of the amplitude R,

$$R = \sqrt{I_s^2 + I_c^2}.$$

In case of no detuning, i.e., $\alpha = 0$, and of small amplitude, the solution of (14) is given simply by

$$I_s = I_{s0} \exp(\pi f(\Gamma - \delta)t)$$
$$I_c = I_{c0} \exp(-\pi(\Gamma + \delta)t) \quad (15)$$

where $\omega = 2\pi f$. Therefore, in case $\Gamma > \delta \geq 0$ holds, the sine component I_s will increase exponentially as described in Section II, while the cosine component I_c decreases exponentially.

The solution of a nonlinear differential equation such as (14) will be presented as integral curves or loci in the (I_s, I_c) plane and the behavior of these curves will be characterized by the singular points, i.e., points in (I_s, I_c) plane where both I_s and I_c vanish (cf. [19], [20]).

The singular points of (14) will be obtained by placing $\dot{I}_s = \dot{I}_c = 0$ into (14) and the result may be classified into three cases 1, 2 and 3 depending on the magnitude of the parameters α, Γ and δ, as shown in Fig. 22. In Fig. 22, the abscissa represents $-\epsilon\alpha$ and the ordinate, δ, where $\epsilon = +1$ if $\beta > 0$ and $\epsilon = -1$ if $\beta > 0$. The characteristic curves which form the boundary lines of the three cases are two half-lines parallel to the α axis and a circle of radius Γ with its center at the origin. These three cases will be characterized by the following features.

Case 1

There are three singular points: One unstable saddle point at the origin $I_s = I_c = 0$, and two stable nodal or spiral points at

$$I_s = \pm \sqrt{\frac{2(\Gamma + \delta)}{3\Gamma|\beta|}}(-\epsilon\alpha + \sqrt{\Gamma^2 - \delta^2})$$

$$I_c = \pm \epsilon \sqrt{\frac{2(\Gamma - \delta)}{3\Gamma|\beta|}}(-\epsilon\alpha + \sqrt{\Gamma^2 - \delta^2}). \quad (16)$$

The integral curves of this case 1 have been shown in Fig. 5 for typical values $\alpha = 0$, $\delta = \Gamma/2$, $\beta < 0$. The existence of two stable states, the exponential build up of the small initial oscillation and all other characteristic features of parametrons described in Sections II and V will be explained by the behaviors of the integral curves of this Case 1.

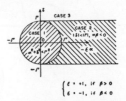

Fig. 22—Classification of the three cases of singular points in (α, δ) plane.

Case 2

There are five singular points: One stable nodal or spiral point at the origin $I_s = I_c = 0$, and two unstable saddle points at

$$I_s = \pm \sqrt{\frac{2(\Gamma + \delta)}{3|\beta|\Gamma}}(-\epsilon\alpha - \sqrt{\Gamma^2 - \delta^2})$$

$$I_c = \mp \epsilon \sqrt{\frac{2(\Gamma - \delta)}{3|\beta|\Gamma}}(-\epsilon\alpha - \sqrt{\Gamma^2 - \delta^2}) \quad (17)$$

and two stable nodal or spiral points at

$$I_s = \pm \sqrt{\frac{2(\Gamma + \delta)}{3|\beta|\Gamma}}(-\epsilon\alpha + \sqrt{\Gamma^2 - \delta^2})$$

$$I_c = \pm \epsilon \sqrt{\frac{2(\Gamma - \delta)}{3|\beta|\Gamma}}(-\epsilon\alpha + \sqrt{\Gamma^2 - \delta^2}). \quad (18)$$

The integral curves of this Case 2 are shown in Fig. 23 for typical values $\delta = \Gamma/2$, $\alpha = 7\Gamma/4$, and $\beta < 0$. In Fig. 23, S, S' indicate the two unstable saddle points and A, A' indicate the two stable spiral points. The presence of the stable saddle point at the origin O indicates that the oscillation is not self-starting. If a suitable initiating voltage is applied to the circuit so as to place the point representing the initial oscillation either in the $+$ region or in the $-$ region of Fig. 23, stationary oscillation respectively represented by point A or A' will be produced. On the other hand, if the point representing the initial oscillation were within the O-region, even a voltage of very large amplitude would not initiate stationary oscillation. As there are three stable states respectively represented by O, A, A' in this Case 2, parametron elements corresponding to this case are usually called "tristable" or "ternary" parametrons, while those corresponding to Case 1 are called "bistable" or "binary" parametrons. In principle, a tristable parametron element may either represent a ternary digit by

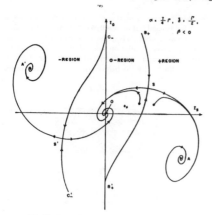

Fig. 23—Integral curves of a tristable parametron.

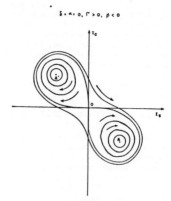

Fig. 24—Integral curves for a loss-free case.

the choice among the three stable states or a binary digit by the choice between the two states, namely, "no oscillation (O)" and "in oscillation (A and A')."

Case 3

There is only one stable singular point at the origin. As the magnitude of the parameters α, Γ and δ are inappropriate, stationary oscillation is not produced in this case.

Now, the functions of the damping will be considered. If there were no damping in (14), i.e., $\delta = 0$, the permissible types (cf. [19], [20]) of singular points will be unstable saddle points and elliptic points, the stability of the latter being neutral. Fig. 24 shows the integral curves for a case in which $\alpha = \delta = 0$, $\Gamma > 0$, $\beta < 0$. O indicates a saddle point at the origin and A, A' indicate two elliptic points. (For a point P on each curve, $AP \cdot A'P = \text{constant}$ is satisfied, and the curves are known as Cassini's ovals.) The point in the (I_s, I_c) plane, representing both the phase and the amplitude of the oscillation, will oscillate indefinitely around the points or point A and/or A', and a generally stationary state of oscillation with a definite amplitude and phase will never be reached. Further, if there were no damping, the oscillation in a parametron would never damp out, even if the parametric excitation were interrupted and it would be impossible to make use of the superregenerative amplification explained in Section II.

Hence, we come to the following conclusion—damping is indispensable both for amplitude stabilization and interruption of parametric oscillation.

On the other hand, if the damping is too large, the building-up rate $\exp(\pi f (\Gamma - \delta))$ of the sine component I_s, given by (15), will become so small as to reduce the speed of the superregenerative action. Therefore, there

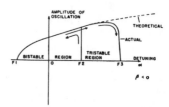

Fig. 25—Amplitude to detuning characteristic of a parametron.

should exist an optimum value of the magnitude of the damping and experimental results show that the optimum value lies in the range $\Gamma/4 < \delta < \Gamma/2$.

Fig. 25 shows a typical example of the amplitude to detuning characteristic of an actual parametron element. In the figure, the abscissa represents the detuning α and the ordinate represents the amplitude of oscillation of a parametron element. In practice, the detuning α may be varied either by varying the tuning capacitance C in Fig. 3(a), or the tuning inductance L in Fig. 3(b), or by varying the frequency or the dc bias of the exciting current in the cases of both Figs. 3(a) and 3(b). The region $F1$ to $F2$ in Fig. 25 corresponds to the above mentioned case 1 and is called "bistable region" since it represents a bistable parametron. The region $F2$ to $F3$, corresponding to Case 2, is called "tristable region," since it represents a tristable parametron. When α is varied continuously a hysteresis jump will occur at the boundary $F2$ between the bistable and tristable region, as indicated by the arrows in Fig. 25. If we assume the presence of nonlinearity only in the detuning as in (11) and (14), the theoretical results indicate that the tristable region should extend indefinitely, as shown by the dotted line in Fig. 25 or by the two half-lines in Fig. 22. Actually, there exists an upper limit $F3$ and this fact will be explained by introducing nonlinearity also in the damping, for example by replacing δ in (11) and (14) by $\delta+\theta I^2$. For bistable parametrons, however, the present analysis assuming the presence of nonlinearity only in the detuning is in good agreement with the experimental facts and generally it is considered satisfactory.

In regard to the nonlinearity of the detuning βI^2, one might think it were caused by saturation of the magnetic cores. If this were the case, β should be positive since the inductance would decrease and the detuning would increase with increasing amplitude. Experiments made on various ferrite and ferroelectric materials, however, show that β is always negative for these materials. On the other hand, it is observed that β is always positive for parametrons using barrier capacitance of semiconductor junctions.

Acknowledgment

The author sincerely wishes to thank Prof. H. Takahasi of The University of Tokyo for his enlightening guidance and encouragement throughout the research. His gratitude is also due to Dr. Z. Kiyasu and the members of the Electrical Communication Laboratory of the Nippon Telegraph and Telephone Company, and to Dr. S. Namba and his associates of the Japan Overseas Telegraph and Telephone Company for their many valuable suggestions and kind advice in the research. He is also indebted to the members of the T.D.K. Company for their cooperation in preparing ferrite material for parametrons.

Bibliography

[1] E. Goto, "The parametron, a New Circuit Component Using Non-linear Reactors," paper of Electronic Computer Tech. Committee, IECEJ; 1954.
[2] E. Goto, "On the application of parametrically excited non-linear resonators," *J. IECEJ*, vol. 38, p. 77; 1955.
[3] H. Takahasi, "Parametron," *Kagaku*, vol. 26, p. 113; 1956.
[4] K. Kamata and F. Sasaki, "Parametron and punched card recorder for the standard meson monitor," *J. Sci. Res. Inst.*, Tokyo, Japan, vol. 51, p. 54; 1957.
[5] Z. Kiyasu, K. Fusimi, K. Yamanaka, and K. Kataoka, "Parametric excitation using barrier capacitor of semi-conductor," *J. IECEJ*, vol. 40, p. 162; 1957.
[6] Y. Nakagome, T. Kanbayasi, and T. Wada, "Parametron Morse 5-unit converter," *J. IECEJ*, vol. 40, p. 974; 1957.
[7] K. Fukui, K. Onose, K. Habara, and M. Kato, "Multi-apertured parametron," *J. IECEJ*, vol. 41, p. 147; 1958.
[8] Z. Kiyasu, S. Sekiguti, and M. Takasima, "Parametric excitation using variable capacitance of ferroelectric materials," *J. IECEJ*, vol. 41, p. 239; 1958.
[9] Z. Kiyasu, K. Fusimi, Y. Aiyama, and K. Yamanaka, "Parametric excitation using selenium rectifier," *J. IECEJ*, vol. 41, p. 786; 1958.
[10] S. Ohshima, H. Enomoto, and S. Watanabe, "Analysis of parametrically excited circuits," *J. IECEJ*, vol. 41, p. 971; 1958.
[11] S. Muroga and K. Takasima, "System and logical design of the parametron computer MUSASINO-1," *J. IECEJ*, vol. 41, p. 1132; 1958.
[12] Z. Kiyasu, "Parametron," *J. IECEJ*, vol. 41, p. 397; 1958. (Circuit Component Issue.)
[13] S. Sajiki and K. Togino, "An Application of Parametrons to Numerical Control of Machine Tools," Rept. of The Government Mechanical Laboratory, Tokyo, Japan; October, 1958.
[14] S. Muroga, "Elementary principle of parametron and its application to digital computers," *Datamation (Research and Engineering)*, vol. 31, September/October, 1958.
[15] Z. Kiyasu, S. Ikeno, S. Katunuma, T. Fukuoka, and K. Hanawa, "An experimental crossbar telephone exchange system using parametrons," *J. IECEJ*, vol. 42, p. 225; 1959.
[16] E. T. Whittaker and G. N. Watson, "Modern Analysis," Cambridge University Press, Cambridge, Eng.; 1935.
[17] N. W. McLachlan, "Theory and Application of Mathieu Functions," Clarendon Press, Oxford, Eng.; 1947.
[18] N. W. McLachlan, "Ordinary Non-linear Differential Equations," Clarendon Press, Oxford, Eng., 1950.
[19] J. J. Stoker, "Non-linear Vibrations," Interscience Pub., Inc., New York, N. Y.; 1950.
[20] N. Minorsky, "Dynamics and Non-linear Mechanics," John Wiley & Sons Inc., New York, N. Y.; 1958.
[21] E. Goto, Japanese Patent 236,746, applied 1954, published 1957.
[22] E. Goto, Japanese Patent 237,701, applied 1954, published 1957.
[23] E. Goto, British Patent 778,883, applied 1954, published 1957.
[24] E. Goto, Japanese Patent 247,326, applied 1955, published 1958.
[25] E. Goto, Japanese Patent 247,327, applied 1955, published 1958.
[26] E. Goto, Japanese Patent 250,529, applied 1955, published 1959.
[27] E. Goto, Japanese Patent Application No. 30-23187, 1955.
[28] E. Goto, Japanese Patent Application No. 30-25583, 1955.
[29] E. Peterson, U. S. Patent 1,884,845, applied 1930, published 1932.
[30] J. von Neumann, U. S. Patent 2,815,488, applied 1954, published 1957. For a discussion of this patent, see R. L. Wigington, "A new concept in computing," *Proc. IRE*, vol. 47, pp. 516–523; April, 1959.
[31] W. T. Clary, U. S. Patent 2,838,687, applied 1955, published 1958.

Esaki Diode High-Speed Logical Circuits*

E. GOTO†, K. MURATA‡, K. NAKAZAWA‡, K. NAKAGAWA†, T. MOTO-OKA∥,
Y. MATSUOKA†, Y. ISHIBASHI†, H. ISHIDA†, T. SOMA†, AND E. WADA†

Summary—Logical circuits using Esaki diodes, and which are based on a principle similar to parametron (subharmonic oscillator element) circuits, are described. Two diodes are used in series to form a basic element called a twin, and a binary digit is represented by the polarity of the potential induced at the middle point of the twin, which is controlled by the majority of input signals applied to the middle point. Unilateral transmission of information in circuits consisting of cascaded twins is achieved by dividing the twins into three groups and by energizing each group one after another in a cyclic manner.

Experimental results with the clock frequency as high as 30 mc are reported. Also, a delay-line dynamic memory and a nondestructive memory in matrix form are discussed.

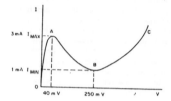

Fig. 1—A typical characteristic of a silicon Esaki diode.

INTRODUCTION

ESAKI DIODES, which are also known as tunnel diodes, are highly suitable elements for logical circuits in view of their extremely high frequency limit, compactness, high stability, and low power consumption. Moreover, the cost can be expected to be very low for mass production quantities in the near future.

An Esaki diode is a two-terminal negative resistance element which is essentially bilateral. Therefore, unlike ordinary transistor switching circuits, Esaki diode circuits require that some special method be incorporated to obtain a unilateral characteristic for the transmission and amplification of digital signals. This situation is completely analogous to that which has been encountered in the case of the parametron.

To illustrate the application of Esaki diodes to logical circuits, a system closely related to the logical principles of parametrons will be discussed in this paper. This system of circuitry for Esaki diodes, based on a proposal by E. Goto, has been developed at the University of Tokyo with the cooperation of the Takahasi Laboratory of the Physics Department, the Amemiya Laboratory of the Applied Physics Department, and the Moto-oka Laboratory of the Electrical Engineering Department. An experimental model whose clock frequency is as high as 30 mc has been successfully built.

THE BASIC PRINCIPLE

A typical voltage-current characteristic of an Esaki diode is illustrated in Fig. 1. It clearly shows the negative resistance region between A and B which is the

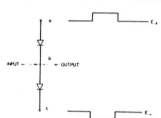

Fig. 2—Basic circuit named "twin."

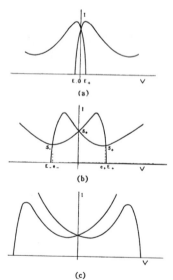

Fig. 3—Response curves of a twin.

---

* Manuscript received by the PGEC, December 15, 1959.
† Dept. of Physics, Faculty of Science, University of Tokyo, Tokyo, Japan.
‡ Dept. of Applied Physics, Faculty of Engineering, University of Tokyo, Tokyo, Japan.
∥ Dept. of Electrical Engineering, Faculty of Engineering, University of Tokyo, Tokyo, Japan.

characteristic feature of Esaki diodes. Two Esaki diodes which have almost the same characteristic are connected in series to form the basic circuit shown in Fig. 2, which will be called a "twin circuit" or more simply a "twin." Symmetric exciting voltages E_+ and E_- of equal magnitude and of opposite polarity are applied between a and c. Depending on the magnitude of these exciting voltages, the twin shows three different kinds of response. When the voltages are small, the operating point of each diode, as illustrated by the two curves shown in Fig. 3(a), lies between O and A on the characteristic curve of Fig. 1. Hence, the potential at the middle point b is zero. Similarly, when the voltages are very large, the operating point of each diode of the twin circuit lies between B and C of the characteristic curve (Fig. 1) and the potential of the middle point b is also zero as shown in Fig. 3(c). When the voltages are chosen so that the operating points of both of the diodes of the twin lie between A and B as shown in Fig. 3(b), there are three possible operating points, S_0, S_+, and S_-. The operating point S_0 corresponds to zero potential at the middle point b, and it is unstable because both diodes are in the negative resistance region. Hence, the operating point of the twin diodes will go to either one of the two stable points, namely, S_+ or S_-, which indicates two possible potentials e_+ and e_- at the middle point b of the twin. These two potentials e_+ and e_- have equal magnitude but opposite polarity. A binary digit can be represented by these two potentials in a twin circuit.

When the exciting voltages are switched from a small value corresponding to the case shown in Fig. 3(a) to a value corresponding to the case in Fig. 3(b), the state S_0 having zero potential at the middle point of the twin will become unstable. The potential must flip to either of the two stable values e_+ or e_- [Fig. 3(b)], and these two should be equally probable for a twin consisting of well matched diodes. Under these circumstances, a very small signal applied at the middle point b will be sufficient to control the choice between the two possible states mentioned above. When the square waves shown in Fig. 4(a) are impressed on the twin as exciting voltages E_+ and E_- together with a small control signal $\pm E_n$, the two permissible voltages e_+ and e_- will build up as shown in Fig. 4(b). This process may be regarded as the amplification of the small input signals $\pm E_n$.

Intercoupling the middle points of the twins with each other by means of coupling resistors, logical operations and the transmission of information will be performed in just the same manner as has been done in parametron (subharmonic oscillator) circuits. Unilateral transmission of information will be accomplished by dividing the twins into three groups, I, II, and III, and exciting each group with one of the exciting signals, $E_{I\pm}$, $E_{II\pm}$, and $E_{III\pm}$, which are switched on and off one after another in a cyclic manner as shown in Fig. 5. The direction of information flow will be from group I to II, II to III, and III to I.

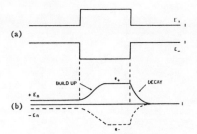

Fig. 4—Switching waveform of a twin.

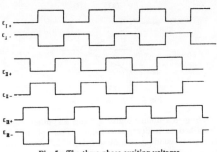

Fig. 5—The three phase exciting voltages.

Majority operations can be performed by a circuit shown in Fig. 6(a). The output of three twins X, Y, and Z in group I is applied to a twin U as its input signal. As the algebraic sum of the three signals from twins X, Y, and Z gives us the effective input of U, the state of U, which represents a binary digit u, is determined by the majority of the three binary digits x, y, and z, represented respectively by the polarity of the potentials of the middle points of twins X, Y, and Z.

Hereafter, in order to simplify the schematic circuit diagrams, we shall use the same conventions as those used for parametron circuits; that is, each twin will be represented by a small circle. Each pair of circles will be connected by a line when corresponding twins are coupled, one line being used per unit coupling intensity. In each circle representing a twin, the input coupling lines will come into the left side of the circle and the output will go out from the right side of the circle. Instead of showing the existing voltages explicitly, Roman numerals I, II and III will be written above the circles to indicate the kind of exciting voltages (cf. Fig. 5) being used.

AND and OR operations can be regarded as special cases of the majority operation with a constant bias. A symbol + will be inscribed in the circle representing a twin to indicate a constant input of unit intensity cor-

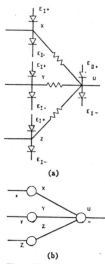

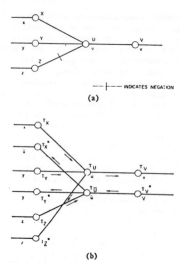

Fig. 6—The majority logic.

Fig. 7—(a) A majority operation with negation.
(b) Negation in symmetric circuitry.

responding to a binary digit "1," and a symbol − will be inscribed to indicate a constant input of unit intensity corresponding to a binary digit "0." Accordingly, a circle with + having two input lines represents an OR circuit and a circle with −, an AND circuit. Using these conventions the schematic diagram of the circuit of Fig. 6(a) is shown in Fig. 6(b).

Negation or the NOT operation is another basic operation indispensable for general purpose logical circuits. In the Esaki diode twin circuits, however, we shall encounter some difficulties in making a NOT circuit. A NOT circuit changes a binary digit "1" into "0" and "0" into "1," and this means the reversal of polarity of dc signals in the Esaki diode twin circuits. In parametron circuits, since the signals are pure alternating currents, the reversal of the polarity is made simply by means of phase reversing transformers. In twin circuitry, since the signals are direct currents, obviously, transformers cannot be used for the reversal of their polarity. Of course, vacuum tube or transistor amplifiers can be used for reversing the polarity. These amplifiers, however, will cause serious signal delay, which makes these amplifiers unfavorable for extremely high speed operation.

In the twin or Esaki diode circuitry, it is possible to make the NOT operation without a delay by employing a rather elaborate system which may be termed the symmetric or push-pull system. In this system, a pair of two twins are used in a push-pull manner so that when one twin in a pair is holding a certain binary variable x, the other twin in the pair holds the complement $\bar{x}$. For example, the logical circuit shown in Fig. 7(a) is a circuit for making the majority u of three variables x, y, and (not z), and for transmitting u to the next stage v. This logical operation will be carried out by the circuit shown in Fig. 7(b), in which the pairs of twins representing binary variables x, y, z, u, and v are denoted by T_X, T_X^*; T_Y, T_Y^*; T_Z, T_Z^*; T_U, T_U^* and T_V, T_V^*. It can be seen from the figure that by setting up the symmetric configuration once at the input, the entire circuitry of a machine can be constructed in the push-pull configuration. Calling the twin in a pair with no asterisk the front (push) twin, and the other (with asterisk) the back (pull) twin, a coupling line without negation in the logical diagram [Fig. 7(a)] may be interpreted to require coupling between corresponding front twins and between corresponding back twins. A coupling line with negation may be interpreted as the cross coupling between the corresponding front twins and the opposite back twins.

The increase of number of elements is obviously a disadvantage of this symmetric system. On the other hand, besides the speeding up of negation, there are two other interesting advantages of the symmetric system. One is its single significant error detecting property. An erroneous operation in either T_Z or T_Z^* in Fig. 7(b) will be called significant if $x=\bar{y}$ and $\bar{x}=y$ hold. The presence of an error in this case will be detected by the fact that $v=\bar{v}$ at the twins T_V and T_V^* in the last stage. On the other hand, if $x=y$ and $\bar{x}=\bar{y}$, the final result will be $v=x=y$ and $\bar{v}=\bar{x}=\bar{y}$ independently of z and $\bar{z}$. Hence, an erroneous operation of twins T_Z and T_Z^* will not have any significance in the result of the computation. Suppose we have a large scale computer made entirely of this symmetric scheme. Then, an erroneous operation of a twin in the accumulator will be significant if it occurred just before printing out of the accumulator content, and it will not be significant if it occurred just before the content is reset to zero. Therefore, it will be possible to detect significant errors by providing a relatively small number of comparators at the output stage of a large scale computer.

The other advantage is that the signal currents will be balanced perfectly in the symmetric system as shown in Fig. 7(b). In very high speed computers, spurious signals induced by common ground currents would be a very serious problem. By placing the twins in each pair closely together in the symmetric system, the ground currents will be balanced out, and the undesirable effects of ground currents can be completely eliminated.

EXPERIMENTAL RESULTS

An experimental model of a binary counter using the symmetric system has been successfully built. The logical diagram of the circuit is shown in Fig. 8(a), and the complete circuit in Fig. 8(b). In Fig. 8(a), 1, 2, and 3 form a flip-flop circuit to absorb the undesirable effects of chattering in the input switch. 4, 5, and 6 form a so-called digital differential circuit and a single pulse is obtained at 6 each time the state of the flip-flop changes from "0" to "1." 7, 8, 9, and 10 form a binary counter. Therefore, each time the input switch is switched from − to + the binary counter changes its state.

In Fig. 8(b), the input is connected to an asymmetric circuit and it is connected into a symmetric form between 1 and 2* and between 4 and 5* by NOT circuits. For the NOT circuit, both transistor amplifiers and transformer circuit shown in Fig. 9 were tested and both operated successfully. In Fig. 9, as the input to the twin in group II is only one, the transformer without dc restration can be used for negation. However, in this case the negation is accompanied by one stage of delay.

The equivalent circuit of an Esaki diode is shown in Fig. 10. The resistance with an arrow represents the dc characteristic shown in Fig. 1; C is the parallel capacitance, and R_s is the series resistance. The maximum switching speed will be determined by the time constant $\tau = C|-r|$, where $|-r|$ is the minimum of the absolute magnitude of the negative resistance.

Using silicon Esaki diodes made by the Sony Corporation, Tokyo, of which the specifications are $I_{max} = 3$ ma, $|-r| = 100$ ohm, $C = 400$ pf, $\tau = 4.10^{-8}$ second, and using coupling resistors of 2000 ohm, the binary counter circuit of Fig. 8(b) has been operated at 1 mc. Similarly, using germanium Esaki diodes (made by the Sony Corporation) of which the specifications are $I_{max} = 3$ ma, $|-r| = 10$ ohm, $C = 40$ pf, $\tau = 4.10^{-10}$ second, and using coupling resistors of 500 ohms, the same counter circuit has been operated successfully at the clock frequency of 30 mc. The frequency was limited to 30 mc because of the characteristic of the oscilloscopes presently available at the University of Tokyo. By comparing the time con-

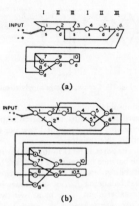

Fig. 8—(a) A binary counter with logical input. s indicates asymmetric circuits. d indicates symmetric circuits. (b) Full circuit diagram of Fig. 8(a).

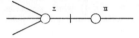

Fig. 9—NOT circuit by transformer.

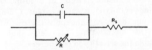

Fig. 10—Equivalent circuit of Esaki diode.

stants τ of both diodes, it seems possible to obtain a clock frequency as high as 100 mc with the present germanium Esaki diodes.

In these experiments the value of the coupling resistances have been determined so as to ensure a logical gain (the maximum number of inputs+outputs) of 10. From these experiments one may observe the fact that the relation between clock frequency f of the twin circuitry and the time constant should be given approximately by $f \cdot \tau \simeq 8$ to 25. This fact implies that the future development of better Esaki diodes having time constants of less than 4.10^{-11} second would result in a billion bit rate (1000-mc clock) machine.

Exciting power supply circuits used in the experiments are shown in Fig. 11. Instead of the square waves shown in Fig. 5, a superposition of dc biasing voltages

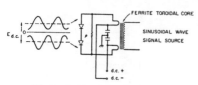

Fig. 11—Exciting voltage supply circuit

and pure sinusoidal voltages is used. The resistor ρ(20 ohms) is used for reducing the source impedance, and most of the source power (about 1 mw per twin circuit) is consumed by this resistor.

One of the important facts in the Esaki diode twin circuitry is the balance between the diodes. The maximum gain and stability depends critically on the balance. It is found that balancing I_{max} of the diodes is most critical. In our experiment, I_{max} in a twin was matched within ±3 per cent tolerance to insure the operation of the counter [Fig. 8(b)] consisting of seventeen twins. The dependence of the balance on the parameters of diodes has been investigated by using the parametron digital computer (PC-1) which simulates the Esaki diode circuitry. The results will be published in the near future.

Memory Circuits

Two kinds of memory devices using Esaki diodes have been tested. The one is a serial delay-line memory proposed by E. Goto. The circuitry of this memory is shown in Fig. 12. A coaxial delay line cable with an open reflecting end is connected to the middle point of a twin. The state of the twin is controlled by the reflected signals, and a circulating dynamic memory circuit is formed. A 16-bit memory circuit at a 30-mc clock frequency using standard coaxial cables (75-ohm impedance, 5-mm diameter, and polyethylene filled) has been operated successfully. This delay line memory will be suitable for serial type computers.

The other is a nondestructive readout matrix array of diode twins proposed by K. Murata. Fig. 13 shows the basic circuit for each binary digit which is inserted at the cross point of an X-Y matrix array. In the normal state, dc holding signals (of value corresponding to Fig. 3(b)) are applied to A_+ and A_-. The nondestructive readout is made by varying the voltage of one of the X lines A_+ and A_- and by sensing the polarity of variation of the current in the Y line or lines. A double coincidence writing is effected by varying the voltage of A_+ and A_- of an X line to facilitate the change of the state

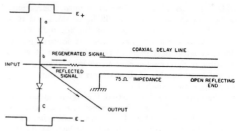

Fig. 12—Delay-line regenerating memory.

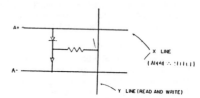

Fig. 13—A twin diode memory cell.

and by applying a writing voltage to a Y line. The results of tests made on a single basic unit are very promising. However, the number of Esaki diodes available at the present has been insufficient to build a full scale experimental matrix.

Acknowledgment

The authors wish to express their sincere gratitude to M. Ibuka (President), and Dr. L. Esaki of the Sony Corporation, Tokyo, for much useful information, and for supplying diodes without which it would have been impossible to accomplish the experiments. Gratitude is also extended to Professor H. Takahashi of the Department of Physics and to Professor A. Amamiya of the Department of Applied Physics of the University of Tokyo for helpful discussions and encouragement.

Bibliography

[1] L. Esaki, "New phenomenon in narrow Ge p-n junctions," *Phys. Rev.*, vol. 109, pp. 603–604; January, 1958.
[2] H. S. Sommers, Jr., "Tunnel diodes as high-frequency devices," Proc. IRE, vol. 47, pp. 1201–1206; July, 1959.
[3] E. Goto, "The parametron, a digital computing element which utilizes parametric oscillation," Proc. IRE, vol. 47, pp. 1304–1316; August, 1959.
[4] E. Goto, and the University of Tokyo V.H.S. Computer Research Group, "On the Possibility of Building a Very High Speed Computer with Esaki Diodes," Electronic Computer Tech. Committee, Tokyo, IECEJ (Japanese); October, 1949.